REALITY AND THE PHYSICIST

Knowledge, duration and the quantum world

By the same author

Conceptions de la physique contemporaine
Hermann, Paris 1965

Conceptual Foundations of Quantum Mechanics
Addison-Wesley, Benjamin, Cummings, Reading Mass.
second ed. revised 1976

Un atome de sagesse
Le Seuil, Paris, 1982

In Search of Reality
Springer, New York, 1983

REALITY AND THE PHYSICIST

Knowledge, duration and the quantum world

BERNARD D'ESPAGNAT

University of Paris

TRANSLATED BY DR J. C. WHITEHOUSE

University of Bradford

AND

BERNARD D'ESPAGNAT

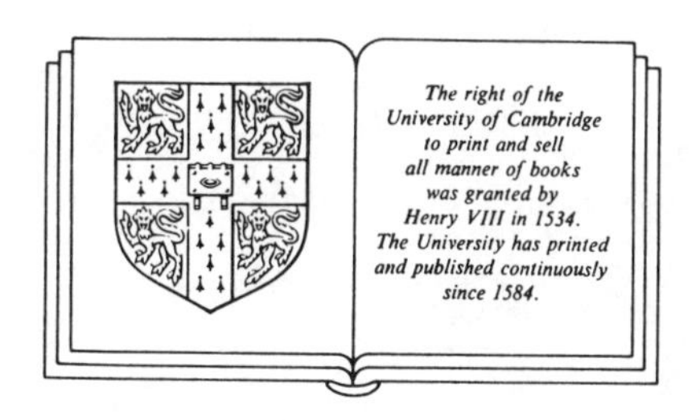

CAMBRIDGE UNIVERSITY PRESS

CAMBRIDGE

NEW YORK PORT CHESTER

MELBOURNE SYDNEY

Published by the Press Syndicate of the University of Cambridge
The Pitt Building, Trumpington Street, Cambridge CB2 1RP
40 West 20th Street, New York, NY 10011, USA
10 Stamford Road, Oakleigh, Melbourne 3166, Australia

Originally published in French as *Une incertaine réalité*

First published in English by Cambridge University Press 1989 as
Reality and the physicist
Reprinted 1990

Printed in Great Britain at
the University Press, Cambridge

British Library cataloguing in publication data

Espagnat, Bernhard d', *1921–*
Reality and the physicist.
1. Physics – Philosophical perspectives
I. Title ex. II. Une incertaine réalité.
English
530'.01

Library of Congress cataloguing in publication data

Espagnat, Bernhard d'.
[Incertaine réalité. English]
Reality and the physicist: knowledge, duration and the quantum world / Bernhard d'Espagnat.
p. cm.
Translation of: Une incertaine réalité.
Bibliography: p.
Includes index.
ISBN 0 521 32940 X. ISBN 0 521 33846 8 (paperback)
1. Reality. 2. Physics–Philosophy. I. Title.
QC6.4.R42E8713 1989
530'.01–dc 19

ISBN 0 521 32940 x hardback
ISBN 0 521 33846 8 paperback

GO

Contents

Preface

We must reckon with two realities. More precisely, present-day physics calls for a clear-cut distinction between two notions both designated in the past by the word 'reality'.

The first of these is the notion of *independent reality*. By definition, it includes everything that *is*; if God exists, or if the world exists by itself, each is 'real' in that sense of the word. This kind of reality, present-day physics indicates, is distant or, indeed, 'veiled'. The second notion is that of empirical reality, the totality of phenomena – a reality which we understand better with each day that passes.

Can we do without either of these notions? Many think we can, and in the past a number of philosophers and physicists have tried to show this is indeed possible. Some of them – the positivists for example, with their much-vaunted rejection of metaphysics – ruled out the idea of independent reality on the grounds that it is meaningless. Others, the materialists or realists, subsumed the notion of empirical reality under that of independent reality, regarding phenomena as the very stuff of all reality. Although both of these schools of thought enjoyed a measure of success, it is my thesis that their day is past and that in our time science itself has provided us with pressing reasons for accepting the (philosophical) duality of Being and of phenomena.

In a simplified and perhaps somewhat distorted form, this is the conception presented here: a conception which present-day physics now presses upon us. How is it connected with the (general) theory of knowledge and with the debate about indeterminism? Or with recent developments having to do with irreversibility, with the highly complex, and with time? How does it bear on the age-old problems, conventionally called 'philosophical', such as the relationship between consciousness and things that can be perceived, or the nature of freedom? These too are questions to be investigated in this book.

With regard to the *methods* by means of which such investigations are to be conducted, they are those of a physicist. This is appropriate since the field under study is one in which a somewhat rare phenomenon, to wit a cross-fertilisation of epistemology and experimental physics, has recently taken place. Indeed, during the 1960s and the 1970s we were but a small band of theorists interested in the problems having to do with the *conceptual* foundations of quantum physics. It is worth noting that in effect the questioning by the various people comprising this band played a major part in having such experiments as those of Clauser, Aspect and others actually performed. And conversely these experiments *did* bring a number of the problems at hand within the realm of experimental physics.

For all that my aim here is of course not to present raw scientific findings. What I am trying to do – using physics as my starting point, but not only physics – is to arrive at an understanding of reality that is in harmony with the present state of our knowledge. Many of the questions that require particular attention are therefore of quite a general epistemological kind. They mainly relate to the quantum nature of fundamental physical laws and also to the nature of complexity and of irreversibility. Though not quite central to my inquiry here, these last two themes are given an important place. In connection with them it is, however, worth bearing in mind that whereas most physicists regard advances in this field as just one part of current progress in physics, it seems that some others see them as constituting a revolution without precedent, one that will dissuade science from continuing its quest for coherence, so that it will 'finally' become receptive to the ideas of *contradiction, uncertainty* and so on. The exaggeration in such a view is plain, not least since operational physics discovered – and, at its level, fully mastered – the interrelated notions of uncertainty and indeterminism as early as the start of the present century. In short, suffice it to say on such matters that it is not *physics* but rather *independent reality* that may legitimately be called uncertain (and that, we should appreciate, in the way that Ronsard speaks of the shadows of the forest as 'uncertain'). But at all events, as we shall see, the advances mentioned above are of considerable significance, once they have been placed in proper context.

In a subject such as the present one, there are endless byways. But here we cannot afford to follow them and to devote attention to problems which are not essential to the understanding we seek to achieve. Consequently, I have striven to give the reader such benefit as I can from my own explorations of such byways and my discovery that they *are* byways; so I have cut out all that I deem not essential to the matter in hand. Nevertheless, the subject remains not only difficult – which is not too discouraging – but also from time to time rather dry – which is more so. For that reason, sections which can be left for reading later have been indicated in the text. In general, to gain a balanced view of the book on first reading, I suggest the following procedure be adopted:

> Part I: read all of Chapter 1, The positivism of the physicists, and preferably Chapter 2 also, together with the first section of Chapter 3;
> Part II: read Chapter 4, Physical realism and fallibilism; sections 1 and 2, and Chapter 5, section 7;
> Part III: read Chapter 9, Independent reality, Chapter 11, Questions and answers, sections 1 and 2, and Chapter 12, Summary and perspectives – in that order, or any other preferred.

The reader who follows this course will, I hope, later want to 'fill in the gaps' by reading the other chapters.

I am pleased to acknowledge Mr Peter Willis' considerable help in preparing the English text.

Introduction

Philosophy is essential. For times immemorial it was used to impart some shape to views that were formless but illuminating, and it continues to do so in our day.

But today as in the past pure philosophy may also delude unwary minds. The unwary tend to have a craving for light and feel they have no time to waste on mere details. Like Descartes shaping his method, they believe (and at the outset, why should they not believe?) that the wings of thought are strong enough to carry them unerringly to the regions where truth prevails and that they have scant need of facts. So they turn to philosophy.

Some indeed are successful in following such a course. But nowadays in the scintillating, esoteric world to be found there many run the risk of getting lost. For *any* highly-polished but opaque way of speaking which sports with meaning and dismisses all that is familiar can easily be mistaken for the pronouncements of ultimate truth. And all too often the false glitter of preciosity beguiles the questing mind.

By contrast, many scientists will accept nothing but facts and 'formalisms' connecting facts. And among them, the physicists (principally the theoretical physicists) distinguish themselves further by their unequalled caution. This caution extends to the distrust – instinctive and acute – of any notion that is neither mathematically nor operationally defined. Even if they know full well that not everything can be defined, they are averse to making use of words from outside that realm; for considering their science, its methods, its progress and its history, they quite properly are on their guard against the shifts of meaning and the misunderstanding of many kinds that the slightest relaxation may let through. But however sound such an attitude is, for many it has the unfortunate effect of diverting them from the great questions, so much so that ultimately, for all the conceptual apparatus we now have at our disposal, it remains difficult to see these issues in their true light. This is all the more so in that there is a well-established tradition in the philosophy of science according to which that discipline should always operate from a position well upstream of current scientific research and restrict itself to reflecting on the general conditions which determine the validity of this research. From this it is plain that a philosophy which is less concerned with prescribing methods to the sciences or to criticising scientific values and more to better understanding the factual results that sciences provides us with is still to be created, or at least stands in need of considerable development.

It is essentially in this spirit that I shall seek in this book a better understanding of what is truly contained in the three great concepts, fundamental for human beings, of reality, causality and time. But if in broad outline this is the aim, achieving it will require a fairly lengthy journey. Such concepts can be, indeed have been, approached from various directions, each corresponding to defensible intellectual positions. These need to be examined without bias. It should come as no surprise to find detailed accounts here of methods and concepts which ultimately do not figure in the conclusions I arrive at. To defer judgement in such fashion clearly would have no place in a work that seeks popular appeal – it would impede the flow – but it is necessary for the balance and proper appreciation of any which, like this one, is an investigation and in which, in consequence, the author must provide precise reasoning in support of his arguments.

Indeed, it should be clear that this book is an investigation and not a popularisation. No judgement of value or of 'status' is implied by that statement. What is meant is simply that in common with any scientific or philosophical article, the book presents the author's own views and not those of 'the scientist' or 'the philosopher', taken generally; it is written with a view to making public some arguments, and ultimately some conclusions, which are not yet part of the corpus of knowledge or of the ideas of specialists. The only difference between it and such articles or other similar studies is one of form. In view of what has already been said, it seemed essential to make the book comprehensible to readers from disciplines other than my own. This has meant adding to the text some reminders of scientific notions that are known in principle but which not all readers may be able to recall immediately. These 'popularist passages' are few and short, and relieved of almost all mathematical apparatus (which however, it is necessary to appreciate, implicitly underlie them). References are given to works in which the mathematics is set out.

It may be objected that a philosophy of science which seeks a better understanding of factual results is primarily a matter for professional philosophers. Are they not in a better position to provide it than a scientist? There can be no clear-cut answer to that question, it seems to me. It is clearly true that philosophers have a great deal to say about the matter, and there is every reason to believe that in the future they will once more provide the major discussions. But it will be a difficult business for them to take over the reins and have everything under their control, for whether we like it or not things are rather more complex and problematic than at one point might have been thought. A clear division of labour, with the physicists dealing with the technical matters and the philosophers with the fundamentals, would work smoothly only if in practice it were possible to isolate the major questions at issue here – reality, causality, the reversibility or irreversibility of time – from matters of pure 'technique', and study them separately, without reference to technicalities. Now there is no logial inconsistency in conceiving a world in which this would be so, but it so happens that such a world is not ours. In the world we live in it has proved impossible to investigate in any valid way general questions we want to study, without referring at some point or other to particular developments which, though technical in structure, have effects that go beyond the mere framework of technique. In other words, alas, such investigations require some acquaintance with theoretical physics.

Some might find this disturbing. Is technical knowledge of theoretical physics also needed if one is to understand the *results* of the investigations recounted in this book? The reader will soon find out for himself that this is not the case (handsome creatures cover their skeletons . . .). Reassurances on another, more subtle point are also in order. Some might have initial trepidations that, like other physicists, I have devised some new formal mathematical structure of physics on an entirely personal basis, and have used it to produce the results in question. If this were indeed the case clearly it would tend to reduce the credibility of a book which purports to say little about the technical apparatus that underlies it. But again it should be appreciated that here things are quite different. What is left implicit in what follows is just the details of an array of formulae and equations which would take up much time and patience to review but which makes up the everyday stuff of present-day physics. It can be found in any textbook of quantum physics, and anyone who wants to do so can look up the details of the mathematical material of which only the substance is indicated here. What is original, or claims to be original, in this book are just the reflections, hypotheses and arguments bearing on the interpretation of this body of public knowledge.

If I had to sum up in a sentence or two the aims of this book, the following would be the least misleading: 'It was once thought that the notion of Being must be repudiated. Now that it has finally become apparent that to do so is to court incoherence, it is dismaying to find that in the interim it has become peculiarly difficult, if the facts are to be respected, to rehabilitate that notion.'

In the present context, 'difficult' does not mean 'impossible', as we shall see. Nevertheless, the summary statement just given may well seem somewhat opaque. It requires elaboration. The rest of this Introduction seeks to do just that. It both summarises some basic facts described in an earlier book* and outlines in preliminary fashion the main ideas of the present one.

To that end, the first step is to pass from 'near realism' to 'mathematical realism'. Niels Bohr described classical mechanics as being 'based on a well-defined use of images and ideas *which refer to events in everyday life*' (my italics). If we bear in mind that the ideas of force, point mass, mass,

* *In Search of Reality* [2], henceforth referred to as ISR and not to be confused with either 2A or 2B; in what follows I use the reference ISR [2] (the numbers in brackets refer to the Bibliography, p.272).

three-dimensional space, and absolute time are the foundations of classical theory, we must grant that his remark is pertinent. Of course there are those who would refuse to accept it, pointing out on the one hand that our everyday experience is that of weight, not that of mass, that of extended objects, not that of point masses, and so on, and on the other hand that classical mechanics includes notions (such as energy and acceleration) that are mental constructs. There is some truth in such observations. But in each case the later notion is defined by means of the former, which in turn are after all very simple idealisations of everyday experience. So if with respect to its fundamental notions we can say that classical mechanics distances itself from naive realism, this distance is rather small. And a realism that is based on the definitions of the kind we have just looked at may well be described as 'near realism'.

If for a moment we leave physics and turn to the 'macroscopic' sciences, it is easy to see that in our own day they are still in practice dominated by near realism. This of course explains why the need to go beyond such realism, though vaguely recognised, is not generally appreciated. Our first task here is to assess this need. It is not difficult. Notions such as 'four-dimensional' or 'curved' space are clearly not simple idealisations of everyday life. Though when we speak of them we use familiar words, all such notions – and more generally, all notions of relativity theory – draw their meaning from mathematical physics. In contrast to classical mechanics, the mathematical physics in which such notions come into play cannot itself be seen as merely a quantitative account of relationships between realities which, qualitatively, could be described in everyday language. To pass from near realism to a purely mathematical realism is thus in fact the first step in a process of abstraction to which physics invites us. As we shall see, mathematical realism has had a great many eminent supporters among the physicists of our age. One of the most ardent of them was Einstein.

Classical mechanics also has a characteristic that seems so self-evident to us that generally we do not even notice it is there. I am referring to the fact that it describes the phenomena it deals with as existing in themselves quite independently of the way in which we come to take cognisance of them, that is to say, quite independently of the experimental apparatus which may or may not be set up to that end. Classical mechanics tells us, for example, that between any two solid objects there exists a force proportional to the inverse square of the distance between them. Now a claim of this kind obviously implies that such a force exists whether we observe it or not, and that therefore if it is observed (if we note the relative acceleration the two objects are subject to, for instance) its various characteristics in no way depend on the instruments used in the observation.

Does relativistic classical mechanics have the same characteristics? This question has led to some argument because Einstein, its creator, from the start gave a major role in his reasoning to the observer. However it seems that relativity, both special and general, can also be seen as describing phenomena ('events') which exist in themselves and which various observers only 'see' – so to speak – from different viewpoints. This at least seems to have been Einstein's own position; it is what enabled him to adhere without contradiction to the thesis of mathematical realism.

And where does quantum mechanics stand in all this? It is well known that Einstein could never accept what is called quantum, as opposed to 'classical', physics. Why was that?

There are some rather simplistic ideas abroad on this matter. They turn on his familiar quip that 'God does not play dice' and consist in the belief that of the characteristics of quantum mechanics quantum indeterminism was the main – perhaps the only – one he refused to accept. Now it is quite true that he was more inclined to the determinist thesis than the indeterminist one. However, a thorough study of his writings shows that a much more profound reason for his reservations lay in the acute tension he soon detected between quantum mechanics on the one hand and the notions, for him interrelated, of reality and localisation on the other. To keep the explanation short, we need merely recall that in Einsteinian relativity the notion of *event* is central. It would not be too crude an over-simplification to say that Einsteinian relativity boils down to combinations of events. Now an event is of course *local*: it is defined in terms of a point in the space-time continuum. The mathematical realism of relativity thus tends to elevate essentially local entities to the status of absolutes (or elements of reality-in-itself), perhaps even conceives of them as the only absolutes. (In this connection it should be borne in mind that carrying out a measurement is also an event or, in other words, is local.)

Now from this point of view, and moreover from that of common sense, quantum physics soon raises serious problem. Here in this Introduction is not the place to enter into any technical arguments. But if, as an example, we take the elementary formulation of quantum mechanics, which is based on the notion of the wave function, we encounter the well-known problem which arises from the fact that it is very hard to avoid using the notion of the *collapse* of a wave function occurring when a measurement is made. If as is often the case the wave function has, before the measurement is made, appreciable values in places distant from each other, the effect of the collapse is generally to change these values immediately and considerably, even at points far from the place where the measurement was actually made. Is this instantaneous non-local effect a physical phenomenon or not? The answer depends on how we interpret the wave function in question – which will be discussed later. Suffice to say at this point that there is no simple way of getting rid of the problem.

Might we not object that Bohr long ago answered all such Einsteinian questioning concerning wave function collapse and related themes? In a way this is so, but his answer includes a vital element without which it would not make sense. This remarkable element is quite simply the abandonment of the idea, which as we have seen is implicit in classical physics, that the various phenomena science describes for us all have an existence in themselves, quite independent of whether and of how they are observed. To give concrete form to his idea, Bohr explicitly proposed that the word 'phenomenon' be severely restricted in meaning and use: that it be used only to refer to 'observations made in well specified circumstances, including the description of the whole experimental set-up' [1].

This proposal of Bohr's has considerable implications. It means, for example, that the quantum systems we call 'particles' (electrons, quarks, etc.) and which are said to constitute objects have no properties (indeed, in relativistic physics, scarcely any existence) *in themselves*. These they have solely *for us*, and this in ways that depend on the kind of instrument by means of which they are observed. Further, in certain cases the experimental apparatus may well include several measuring devices set up at whatever distance from each other we may desire. In such cases, Bohr's proposal clearly implies some sort of evacuation of the meaning of the world 'local'. As we can see, it is a long way from the mathematical realism of Einstein.

For a long time, physicists gave up the idea of digging more deeply into such questions, finding that in their eyes they had taken too philosophical a turn. For them there was, and still is, a more urgent need: to make use of the calculation rules of quantum physics in the very many domains where prodigious advances were and still are possible thanks to that wonderful tool. Many of them, moreover, were developing a conviction which though often tacit was none the less genuine: that it is still possible to believe in locality (which if abandoned in certain cases has extraordinary consequences as regards admissible world-pictures) provided only that it is not to be introduced into any reasoning process in quantum physics. Since it could not be put to any use, the idea remained for the most part unformulated. If it had been formulated however, it could well have been couched in terms that could be outlined as follows: 'On the one hand there is quantum mechanics, which provides calculation rules applicable to empirical data and of undoubted validity but which if interpreted in too realist a way are non-local. On the other hand there is reality-in-itself, which could well be composed of some subquantum medium at present inaccessible to experiment but which it is not impossible to conceive as localised.' Such an idea ultimately permitted retention of a world-picture which conformed to the great post-Galilean scientific options (mechanism) without being too much perturbed by the truly strange ideas of Niels Bohr.

Now, as we shall see, an intellectual 'fall-back position' of that kind is no longer tenable. Bell's theorem establishes beyond all possible doubt that any (mathematical) realism of which it is required that it should not contradict the verifiable (in principle at least) and unquestioned predictions of quantum mechanics must necessarily be non-local. This is because of the following reason. If the locality hypothesis were right some correlations at a distance observed in certain circumstances between simultaneous events could only be due to common causes differing from one case to another. Now, for some such correlations the theorem in question shows that if they *were* due to common causes, certain inequalities between measurable numbers would necessarily hold true. However, according *both* to the predictions based on the rules of quantum mechanics *and* to the experimental findings these inequalities are violated.

If the 'common cause' explanation is ruled out, to what can we attribute the observed correlations? The only way out of the predicament is to assume instantaneous transmission of influences over long distances. This means that a consistent realist who believes in the existence of an external reality embedded in space-time can picture it only as one in which such instantaneous influences at a distance do in some circumstances occur. In other words, he must picture it as violating the laws of relativity.

This is the fact of the matter. But here an extremely important point must be borne in mind. It can be shown that such 'relativity-violating' influences at a distance cannot serve to transmit any utilisable signal. In other terms, they are, in that sense at least, inobservable.

This point is in fact important from two points of view. First, it 'rehabilitates' relativity, so to speak – at least as an operational theory: anyone taking into account only 'phenomena' in the restricted sense Bohr gave the term can and must retain the complete relativistic formalism as it stands. Second, and more importantly, it shows that a careful distinction must be made between the *real* and the *observable*. Unless we discard altogether the very idea of a reality that is independent of our knowledge, we have to accept that such a reality cannot be identified with the ensemble of phenomena. This in turn means that we cannot escape what I claim is the fundamental distinction between reality *in itself* or *as such* – reality independent of human minds – and the ensemble of phenomena – or *empirical* reality. As we shall see, this distinction is not for the use of philosophers alone. Scientists seeking to understand in depth the nature of certain debates internal to the scientific community will also find it useful.

Although conducted only in summary fashion, with elaborations left till later, the preceding discussion should already serve to suggest that the distinction just drawn has important implications for the idea of cause and the related idea of explanation. It will also become apparent that it has consequences for the problem of complexity and the closely-connected matter of time.

As regards cause, in the eighteenth century the philosopher David Hume pointed to the conceptual difficulties surrounding the notion; since then the questions raised about it have always been acute. On some occasions it was rejected in favour of the notion of law, on others it played (and still plays) a central role in extremely precise physico-technical developments such as signal theory and dispersion relations. In our own day as in the past, it raises conceptual problems that go beyond the merely technical, questions such as 'should we or should we not distinguish between the idea of causation and that of explanation?' We shall see what light can be thrown on such problems by the distinction between independent and empirical reality. The whole business, it will emerge, has considerable ramifications.

As regards time, a very old and very difficult problem came recently to the fore once more as a result of great progress made in our understanding of complex out-of-equilibrium phenomena. Is irreversible time, the time in which, so we feel, we actually 'live', a reality or an illusion? (We know that the time which figures in the equations of Newtonian mechanics, electrodynamics and so on is 'reversible'.) Here to distinguish between the two realities adds new and it would seem truly significant elements to the discussion. Indeed, these elements cast light not only on the questions we have mentioned but on others as well, such as those of consciousness and of freedom.

A few further remarks should help set the tone of the book. One is by way of a warning. It is important to avoid believing that questions which are in fact closed are somehow open. The seductiveness of what glitters, referred to briefly above, sometimes leads us to see the world as younger than it really is. We may imagine that to reach the truth we only need to come up with brilliant ideas, or at very least (since 'reaching the truth' is after all a rather ambitious aim) that we need only think brilliantly for our thoughts to be valuable and to open fruitful possibilities. In some respects, such an illusion is beneficial, since it guards against discouragement and it maintains some kind of a link between the public at large and genuinely creative thinking. In countries where the illusion is less powerful, those who contribute to the advance of knowledge tend to be even more cut off from the wider world, since what they have to say is seen as too difficult to be accessible to those who are not experts. This leads to a rather sad divorce between silent experts and talkative communicators which at least in this country (France) is rendered less acute by the quaint notion that glitter can be productive (for this notion increases the credibility of the view that even laymen are capable of understanding and appraising arguments on questions of a conceptual kind). But in effect it remains illusory to hope that in our day people can still make valid claims on matters such as reality, time and causality if these claims are not rooted in the extraordinarily elaborate factual knowledge now at our disposal.

A second remark is of quite a different order. Non-scientists whose work brings them into day-to-day contact with scientists often react with mixed feelings to the claims of these scientists. Once the non-scientists have succeeded to some extent in explaining what they have in mind, it becomes apparent that what they are experiencing is a want of intellectual satisfaction. The burden of what they have to say is that while they accept that credence should not be given to the charlatans and false prophets who dress physics in bright colours, and put into its mouth any words they like, at the same time they feel that serious scientists – at least in their scientific discourse – have stripped the world of all taste, smell and succulence. Yet they *know*, directly and therefore authentically, that this is indeed a world full of colour and of solid, subtle and disturbing realities which earlier philosophies, to say nothing of religions, managed to incorporate into their systems of thought. How is it, they ask, that scientists who claim to have taken over from such systems, are no longer able to do so and, what is worse, why do they no longer even understand the general sense of such a question?

To such anxious questioning one may conceive *a priori* of at least three kinds of response. The first would consist, in outline, of claiming that immediate knowledge, or what is said to be immediate knowledge and displays all the appearances of it, may well be nevertheless quite inauthentic; the human mind can create spontaneously concepts that are false in that they are quite ill-fitted to the description of reality. *A posteriori*, science will correct such things . . . A second, less discouraging kind of response is to stress that science is not yct complete and that its findings will not remain eternally bleak. Optimists may consider that the recent advances in the field of complexity open the way to developments along those lines. To borrow an idea from mathematics, we could say that the recently-created concept of *fractals*, which enable mathematicians to model forms such as branches, shores or clouds, provides some notion of what might happen there. And it is also possible to imagine a third kind of response, by no means incompatible with those mentioned so far, which would emerge if it seemed likely – *a posteriori* of course, after close study of the facts and the rules linking them – that the whole of communicable experience (which constitutes the object of science) cannot be equated with the whole of what *is*. The distinction between reality-in-itself and empirical reality clearly opens new perspectives in that direction.

A final remark should help explain why the text includes, on certain questions, elaborations that might seem pointless or needlessly detailed. Talking to a number of people about the problems tackled here soon reveals that most of them have adopted one of two possible but irreconcilable *a priori* positions, which I call the *physicalist* and the *mentalist* stances. Those adopting the first see the existence of a knowable independent reality – which it is the specific task of science to acquaint us with – as a primary self-evident truth whose renunciation would in their eyes render us guilty of a kind of intellectual abdication. For those who adopt the second position – who are almost as numerous as those adopting the physicalist stance – the primary self-evident truth is by contrast the view that any scientific concept is a construct. Human beings can speak only of what they feel, observe or do – in short, of what they experience; so their science can only be an account of that experience, in the broad sense. Both theses are defensible. What is irritating is, as noted, that all too often they operate as unspoken *a priori* assumptions, the intellectual equivalent of taboos, in that they induce immediate rejection of any argument not built from the beginning on whichever of the two positions the person concerned has adopted. It is because of the fact that I have observed most people adopting one or other of these positions, and so have been helped to move beyond both as such, that I find it necessary to spell out some arguments which could (wrongly, in my view) be considered obvious by those who unconsciously share either one of these assumptions.

There are three parts to this book. Part I, Chapters 1–3, deals with what could be called 'the philosophy of experience', or 'the positivism of the physicists', or even 'instrumentalism', if by that word we understand the view that theory (possibly in a highly mathematical form) is essentially an instrument for predicting the results of empirical investigations. Such a philosophy has attracted a great deal of criticism, some justified, some not. It is important to examine it thoroughly right at the beginning, for whether we deplore it or delight in it, the fact is that this philosophy is basic to all current work in theoretical physics.

Part II, Chapters 4–6, is devoted to realism, more precisely to the conception I call 'physical' or 'mathematical' realism. This is a modern version of what in traditional philosophy is sometimes called transcendental realism – the view that there is no essential distinction to be drawn between reality in itself and the ensemble of phenomena. After a period of eclipse, this view returned to favour after the Second World War, at least in the epistemological circles where Karl Popper's influence was dominant. When considered in the light of recent findings in physics, this kind of realism proves difficult to sustain, even in its fallibilist form. So it should come as no surprise that certain chapters in this part of the book, notably Chapter 6, are rather negative in tone (they can be omitted on first reading). It is only when one has thought one's way to the conclusion that a particular idea, though quite natural or seductive in appearance, is in fact untenable that one becomes disposed to take seriously certain other ideas at first glance less plausible but for that very reason capable, if they turn out to be sound, of opening up new horizons.

Part III, Chapters 7–12, is concerned specifically with these new horizons. It examines the problems associated with complexity and the irreversibility of time, seen in the light of the findings of Parts I and II.

Some rather bold ideas of certain contemporary physicists, which have led them (via fully-rational, rigorously argued paths, be it noted) to extreme and mutually incompatible conclusions, are explored in some detail. An extended series of diverse questions (on science and metaphysics, rationality, universality, and the like) are then asked and answered, and finally some insights opened by the new perspectives are suggested.

In approaching the matters raised in the final chapters, extreme caution is advisable. Facts and the principles linking them together are our sheet anchor in such uncertain waters and should never be abandoned. Fortunately, keeping scrupulously to that rule not only does not stop the flow of ideas but even brings them – as we shall see – flooding forward in a way that would be regrettable to hold back. (This last part is where new vistas are mostly opened out; the reader particularly interested in these may like to read it first.)

A note for theoretical physicists

This book, it will have become apparent, is not intended merely for specialists outside my own field. It is also addressed to those in the domain in which I work, theoretical physics. Naturally I would like it to help colleagues in that field better to define the particular concerns of their discipline, perhaps suggesting new lines of research using the techniques of that discipline. I must however point out that it is precisely at such a technical level that I had to compromise here. Every reader, whatever his or her speciality, had to be introduced to the essential features that, in the basic substance of our discipline lead to well-formulated and philosophically interesting problems – and make it possible for us to examine them fruitfully. The book had to be directed firmly towards the discussion of these philosophical aspects, for that is why it was written. That being so, a systematic and exhaustive examination of all the detailed technical questions which in the general opinion of specialists, myself included, can legitimately be asked about the use of the various algorithms and procedures referred to in the text would have been out of place. The book would unavoidably have become very long, would have bristled with technicalities, and would not only have discouraged specialists in other fields but also run the risk of obscuring the real problems even from theoretical physicists . . .

The compromise I have been obliged to make is to refer readers who are theorists to other writings of mine in which the technical details in question (which have not, of course, been lost from sight) are treated thoroughly. The main ones [2] are a long article in the journal *Physics Reports* (henceforth referred to as [2A]) written specifically for this purpose, and an earlier book, *Conceptual Foundations of Quantum Mechanics* (henceforth referred to as [2B]). Any theoretical physicist, novice or veteran, who contemplates engaging in either physical analysis or discussion, or use in his own work of those parts of this book in which such technical details are referred to, is invited, lest his efforts risk coming to naught through lack of firm roots, to supplement the present work by those others, together with the references they include.

This suggestion, let me repeat, is intended only for theoretical physicists. It must be stressed that this book is meant to be self-standing as far as any analysis or discussion of all matters not connected with the technical details of the algorithms used in theoretical physics is concerned. In other words, its content is accessible to any enquiring reader who, with regard to such technicalities, is willing to follow the usual practice of trusting in the good judgement of the professional.

Part I

Instrumentalism and science

I
The positivism of the physicists

Nowadays, chemists would be among the first to acknowledge that it is in the field of physics, and more precisely of quantum physics, that the theoretical and conceptual foundations of their discipline are to be found. Similarly, most astronomers now see themselves as astrophysicists, and with good reason. It would be easy to make further observations of such a kind. In this sense the unification of the empirical sciences, long seen as an ultimate goal, is well on the way, under the leadership of physics.

Given this state of affairs, it is all the more necessary to reflect on the nature of the methods used by physicists and the degree of certainty of the results they produce. Such an enquiry can be conducted from the outside, so to speak. That is what philosophers of science do when they examine contemporary physics. It can also be conducted from the inside, that is to say, by physicists. In view of the complexity of the task, it may well be a good idea for people from both disciplines to undertake it.

The early part of this book is therefore devoted to a detailed account of the physicist's way of seeing things. Or rather, to be more precise, I try there to describe and seek to explain what seems to me to be the way most of my colleagues see them. Once described in what I take to be the proper and necessary way, this way of seeing things is clearly at odds with certain claims currently put forward by some philosophers of science and backed up by arguments that need to be taken seriously. Such a conflict requires a full discussion, in which, however, I shall try to avoid both learned and historical digressions that are too long, and over-simplifications that are too gross.

1 A critique of concepts

When I speak of 'the positivism of the physicists' I mean specifically their way of seeing things, as mentioned above. Ignoring one or two qualifications that need not be considered in a first approximation, we may also use the expression 'the philosophy of experience', to which I had recourse in ISR [2].

From a methodological point of view, the positivism of the physicists has obvious affinities with positivism in the traditional sense of the term, as used to indicate the approach of the philosophers who developed that particular doctrine. In certain major aspects however, the physicist's positivism is so distinct from the positivism of philosophers that it would be wrong to think in terms of some pure and simple borrowing by physicists from philosophers. Indeed, the chief feature of the positivism of the physicists is a characteristic that in traditional positivism is present only in the background, and in the work of some of its proponents, is entirely absent. What I am referring to is a *critique of concepts*, which is indeed of the very essence of the positivism of physicists. If we are to gain any understanding of the physicist's way of seeing things it is vital to grasp the nature of this critique and to understand fully why it is necessary.

In such an investigation we should of course bear in mind, to begin with, that the notion of a concept has a long and complicated history. But it is not the old quarrels between nominalists and universalists that we should dwell on here. When all those arguments were taking place, the field of objective knowledge merely covered astronomical observations and their theoretical synthesis on the one hand, and factual data about material objects on the other. *Any* concept (think for example of the concept 'cathedral') was therefore conceived as being essentially a selection of *some* (immediately given) empirical data, with other data deliberately left aside. In those days, the argument was merely about whether concepts in this sense – the concept of a 'material object' for example, or of 'good' – referred to entities existing in their own right, or whether they were mere names created by human beings for their own use. Whatever the case, it was felt to be fairly certain that such concepts were suitable for whichever objective description seemed to call for them.

As a result of a tacit (and though quite natural certainly unjustified) extrapolation of implicit convictions of this kind, 'reasonable' men (as opposed to dreamers or charlatans) shared for a long time the view that the basic concepts of everyday experience, such as space, time and material object, were applicable not only to the description of any conceivable experience but also to the theoretical synthesis of experience in general, even if it then proved useful to supplement them with certain *derived* concepts such as energy. Until what is in fact the fairly recent past, this way of seeing things, which I have called 'near realism', was shared by a great number of physicists, at least as a first approximation. Should we criticise them for this? Certainly not, for any intellectual advance entails working hypotheses, and in times past such a way of seeing things was undeniably quite natural and so, as a working hypothesis irreproachable.

Times have changed, however. For example, we now know for sure that the concept of time as totally distinct from space is no longer fully valid, no more than is that of space when conceived as having nothing in common with time. Such concepts, which may be useful when we are solely concerned with accounting for customary experience and objects moving at speeds that are low when compared to that of light, are of no help in describing the ultra-high speeds attained by objects of extremely low mass. In such cases, as is now familiar, we must replace them with the concept of 'space-time'. That concept, however, involves a specific rejection of the traditional, intuitive idea that space and time are totally separate and can in no respect be transformed into one another.

Can we object to this kind of relativising of time and space? It is indeed possible, for physics is not dogmatic, but only on condition that we posit as a principle that of all the phenomenologically-equivalent frames of reference, one and only one is 'really at rest'. But then we are faced with the question of why that one rather than any other.[1] Indeed, we are faced with the very question of how to *define* the frame of reference we particularise in this way. No experiment enables us to do that on the basis of general laws. Under these conditions, it is quite obvious that to posit a particular frame of reference as the one at rest would be quite arbitrary, just as deciding to posit the 'verticality' or 'non-verticality' of a dipolar molecule in interstellar space would be. The absurdity of both is plain. And if we try to explain the origin of this absurdity, we shall see that it lies entirely in the fact that there is no way of selecting one particular frame of reference by means of laboratory experiments.

From an historical point of view, great importance must clearly be attached to the discovery that the old concepts of time and space are inadequate in respect of the theoretical synthesis of communicable human experience. Previously, as we said, these concepts appeared self-evident. It seemed obvious that any description of whatever kind could and should be made in terms of them. Once they proved to be inadequate outside a certain experimental domain, *all* concepts therefore came to be viewed with suspicion; or more precisely, it became necessary to question the appropriateness of each of them – even the most 'obvious' – as a constitutive element of a theoretical description that aimed at going beyond the framework provided by our ancestral or familiar experience. Notions previously considered 'clear and distinct', and essential for all thought now appeared to be no more than 'constructs' elaborated either by every young child or by our most distant ancestors (opinions may differ on this). In other words these notions appeared to be just useful components in simple and usually implicit theoretical syntheses of a very limited day-to-day experience. They were therefore seen as notions that could well be inadequate as regards the much wider-ranging called for by our expanding knowledge of all phenomena in the fields of the very large and the very small.

And yet we need words to express what we want to say, and common names – or in other words, concepts – to formulate general laws. In order to make valid statements about the laws governing stars or atoms – the vast and the minute – we must admittedly beware of the ideas we already have at hand (for the reasons just examined), refusing to accept them as they are without thoroughly testing them. But at the same time we plainly need *concepts*. What, then, should they be based on? Since our theories, even those concerned with quasars and quarks, must ultimately be in essence syntheses of our *experience*, or in other words, syntheses of the measurements we have made or shall make, the answer is clear. The words and concepts we use must be defined on the basis of our experience. More precisely, they must refer to our measurements. To put it another way, we shall have to base our language on those terms which refer to what we have *defined operationally*, stating for example that 'by definition, the object under consideration has property *x* if, when acted on in manner *y*, the result *z* is observed'. As is familiar, that is precisely how Einstein, at the dawn of the conceptual revolution he did most to bring about, felt it vital to define the concepts of length and temporal duration involved in his theory of relativity.

The notion of operational definition thus lies at the heart of contemporary physics. The fact that it does so poses serious problems, for we shall see as we proceed that when analysed in detail it raises difficult questions. But the fact itself is beyond dispute. It is true that commentators who are not physicists sometimes forget it, and so go on to propose natural philosophies that are constructed on the basis of the old concepts of space, time and matter conceived of as localised. But however attractive these philosophies may seem, they are nevertheless ineffectual. It must be stressed again that, since concepts once universally considered to be valid everywhere and always have been found wanting, the operational method just described, which makes it possible to give unambiguous meanings to the terms used in scientific descriptions, is indispensable.

This being so, we should note at once that there are at least two ways of interpreting the method. This is an important point, and later we shall return to it at length; but it must be brought to attention as early as possible. One of these interpretations perhaps goes back to Galileo, at least if we are to believe his best-known writings. Indeed it was this pioneer of modern science who wrote that 'the book of the Universe is written in the language of mathematics, the symbols of which are triangles, circles and other geometrical figures, without which it is impossible to understand a single word of it'. If we remember that Galileo, unlike the Greeks, based his science on experimental evidence, we may understand these words to mean that the experiments conducted by the physicist enable him to go beyond appearances and approximate or false notions, and to discover the *true* concepts, that is to say, the words of the language in which the book of the Universe is *actually* written. In any case (and quite independently of any attempt to elucidate Galileo's thought), now that the vulnerability of concepts drawn form everyday life has been established, this way of seeing things *a priori* constitutes a very coherent interpretation of the operational method we have been discussing. To summarise: this interpretation holds that the systematic use of operational definitions and their concomitant incorporation in a mathematically rigorous theoretical framework should eventually make it possible to discover the true concepts that express adequately the very structures of reality in itself: space-time rather than space and time, curved rather than Euclidean space, and so on. In what follows, this interpretation will be referred to either as *mathematical* realism, to remind us that these 'true' concepts are generally mathematical in nature, or as *physical* realism, to indicate that we are dealing with a conception which assumes that physics can legitimately aspire to describe reality 'in itself' – that is to say, a reality quite independent of the human mind – and to describe it, ultimately, *as it is*.

Clearly, an 'ontological' interpretation of this kind is quite compatible with a positivist, or if the term is preferred, an operationalist *methodology*. So there is nothing surprising in the fact that it was maintained by the very physicist – Einstein – who previously had given operationalism the ultimate seal of intellectual respectability. However we must be quite clear that this 'ontological' interpretation is not the only conceivable one. Another, considerably less ambitious, is based on an observation which is quite explicit in Kant's writings but is readily accepted by physicists who are not systematically Kantian in outlook. It is simply that most experiments have a purpose. This does not of course mean that physicists prejudge their results. But they do at least regulate the experimental conditions very carefully, with the aim of making the results intelligible. But intelligible to whom? Not to themselves alone, naturally, nor even to those belonging to their socio-cultural group alone; rather, to the community of human beings endowed with reason. This can only mean that in some sense they force nature to express herself in the language of reason, that is to say, in *human* language.

He who reflects along these lines readily comes to the conclusion that it is definitely over-ambitious to aim at scientific knowledge of the very structures of independent reality. It is true that, recognising the mistakes in Kant's approach to the notions of space and time, he will not go so far as to claim that the language in question is totally *a priori* in the sense that it is settled once and for all. To do that would be to bring back in a more elaborate way the old intuitive notion of clear and distinct ideas, seen both as immutable and as the only valid means of describing the total theoretical synthesis. But in adopting quite unreservedly the operational method of selecting concepts and of attributing precise and unambiguous meanings to them, he will legitimately view this method as an effective procedure for constructing a science which is to be considered only as a synthesis of communicable human experience. This means he cannot assume that scientific knowledge has any significance outside the limits of such experience. Clearly, according to this view of the matter, physical theories can scarcely be seen as anything other than *instruments*, which enable us, on the basis of observed facts, to predict either with certainty or probabilistically the results of observation. This is why this view is often called 'instrumentalism'[2]. Clearly it can also be called 'the positivism of the physicists', 'the philosophy of experience', and other such names.

The restrictive sense in which 'instrumentalism', 'positivism' and the like *clash* with expressions such as 'physical realism' or 'mathematical realism' is not however the only sense they can legitimately be given[3]. It is true that 'in their heart of hearts' physicists currently at work belong to one of two groups: those who support the notion of physical realism; and those who support instrumentalism in the restricted sense described above. However such a difference in outlook, though important at the level of ideas, is negligible in practice. In their professional activities, all practitioners acknowledge the validity of the positivism of the physicists, that is to say, of instrumentalism, *at least as a methodology*. When, as in the discussion which follows, we refer only to methods and the greater or lesser degree of certainty of the results they lead to, we shall be using 'instrumentalism' or 'the positivism of the physicists' to mean purely and simply a methodology. In this sense of the terms, there is no basic opposition to instrumentalism on the part of physicists who endorse the notion of physical realism. Though in their eyes it is insufficient, they nevertheless see it as an indispensable method that in its own sphere is quite trustworthy.

2 A first discussion of instrumentalism

As has just been said, with regard to methods at least, instrumentalism – the 'positivism of the physicists' – is the foundation stone of contemporary physics. Given that this part of science has been astonishingly successful in various fields, from elementary particles to astrophysics, quantum optics, solid-state physics, thermodynamics and chemistry, it is hardly surprising that physicists should see this foundation stone as being very solid and providing a basis for objective certainty. However, a number of philosophers of science are not of the same opinion. Although their doubts have so far not spread to engineering and technological circles, they have certainly been favourably received in the human sciences.[4]

For reasons that will gradually emerge as this work progresses, I too am far from being satisfied at every level with the positivism of the physicists. Taking such a stance however does not require that every criticism of that positivism must automatically be accepted. Rather, what does seem clear to me is that these criticisms themselves should be rigorously examined. This holds true particularly in regard to the *a priori* objection of those who claim that instrumentalism is purely and simply a negation of the cognitive power of science, and hence condemn it out of hand. Even at the most elementary level, knowing that striking two flints together can produce fire is genuine knowledge (although it makes no less pertinent the remark of the realist that, for example, knowing how to drive cars and knowing how their engines work are items of knowledge of quite different kinds). Generally speaking, criticisms of instrumentalism all need to be analysed impartially, for only if we do so can we form an appreciation both of their relative weight and of what remains valid in the positivism of the physicists.

Some of these criticisms relate to practical matters; others are more philosophical and are concerned with the foundations of the theory of knowledge. We need to distinguish clearly between these two categories, for a purely philosophical objection which affects matters of principle only is sometimes wrongly seen as a difficulty which might lead to scientific errors. The sections that follow look at criticisms which are or which claim to be primarily of a practical nature (either because they relate to the fruitfulness of science or because they are concerned with the contents of what it has to tell us). As for the more philosophical criticisms, we shall examine them only after we have reviewed, schematically at least, the conceptual bases of empiricism and positivism, and of the main doctrines which have recently grown up in opposition to them.

First criticism: instrumentalism is sterile

Instrumentalism, we said, ultimately sees physical theories essentially as instruments or rules for calculation, which enable us to use facts already observed to predict, with certainty or probabilistically, the results of observations yet to be carried out.

It is worth making an aside here. Perhaps in order to discredit this instrumentalist conception, it has been claimed in some quarters that it is religious in origin. Its founder is said to have been Cardinal Bellarmino who, it is well known, tried at one point to save Galileo from the wrath of the Apostolic See by urging him to present his theory – or rather that of Copernicus – as a useful rule for calculation and not as a thesis concerned with reality. Of course this is not a criticism many authors would suggest be taken seriously. On the one hand, attacks of that kind can be directed against theories of any kind (as in the opposite situation, where some 'refute' deterministic realism by arguing that Newton may have found the idea for it in the concept of divine law). On the other hand, it is clearly irrational to assess the truth or falsity of a theory on the basis of how much we like or dislike those who hold it.

Much more important, of course, is the accusation that instrumentalism is unproductive. This claim is made principally by 'fallibilists', philosophers of science (about whom there will be a fair amount to say later) who do not merely stress the need for a theory to have some experimental bearing but also join Popper in maintaining that the only cases when we can reach certainty are those in which we falsify a theory, not those in which we strive to verify one. In the view of such people, an instrumentalist interpretation of a theory implies that it cannot be refuted and since – they claim – a process of eliminating theories through successive refutations is the driving force in the development of science, the irrefutability associated with instrumentalism acts as a brake on scientific progress.

Looked at from a certain angle, there seems to be a case to be made for this line of argument. If for example a present-day physicist refuses to declare that Newtonian mechanics is false, and asserts that on the contrary it is correct wherever its concepts are applicable, the fallibilist may initially see such an attitude as a danger to science. He may even judge that the physicist in question is unlikely to be creative, bent as he is on defending ideas that have already been refuted and therefore unlikely to question old theories and look for new ones. Looking upon theoretical instruments in the same way as he does the tools used by a carpenter or stonemason, the fallibilist may also feel that to use them could lead only to the prediction of events of a given type and, ultimately, of a *known* type . . .

It is quite possible that a philosopher of science with fallibilist leanings might, in the solitude of his study, think in such fashion. But if by contrast he decides to take a look at the true history of the development of contemporary physics, he may well be led to wonder whether his analysis has not neglected significant elements. He may, for example, discover that the person making the claim about Newtonian physics just mentioned was none other than the physicist Werner Heisenberg, who in making it showed that he subscribed, to some extent at least, to a form of instrumentalism that is not of a purely methodological nature. Logically, the analysis we are looking at would lead us *a priori* to conclude that Heisenberg's contribution to physics could only be particularly timid, uninventive and insignificant. In fact nothing could be further from the truth: in most domains, contemporary physics of an advanced kind is founded at least in part on ideas put forward by him. So psychologically speaking, the situation is clearly not as simple as it seems (see the remark at the end of this chapter), and it appears that there is something erroneous in the inferences suggested by blanket criticism of instrumentalism. More generally, the whole of the recent so-called 'elementary-particle' physics was built on partly instrumentalist foundations, and still, over the last twenty years there has certainly been no shortage of perfectly testable, and indeed tested, models in that field. This is a clear illustration of the fact that instrumentalism in no way prevents the application of the method of progress by refutation so rightly advocated by Popper and his followers. Moreover, the momentous consequences this kind of physics has had as regards – for example – our knowledge of the structure and evolution of stars, the explanation of pulsars, the 'Big Bang' theory, the invention of lasers, and so on are enough to show that a physics firmly rooted in instrumentalism is quite capable of providing new and interesting predictions in areas other than that in which it originated.

Moreover it is easy to explain why fallibilist reservations have proved in this instance to be of so little weight. In the first place, instrumentalists certainly do not give all theories the same status. They are fully aware that the domain of validity of say Newtonian theory is merely a small 'zone' within that of Einsteinian theory; that is why, although they do not regard the former as false, they nevertheless consider it to be in some degree 'extinct'. However it should be noted that for instrumentalists such evaluations are always qualified. Thus for example they do not regard the wave theory of light expressed in Maxwell's equations as 'extinct' in any sense of the word. Although the domain of validity of that theory is in a certain way covered by that of modern quantum electrodynamics, it (and what goes with it) is in fact still indispensable not only for the technical calculations but also in a sense for the very formulation of quantum electrodynamics, albeit that the latter theory goes beyond it. More generally, *the equations remain*.

Secondly – and this is a reason which touches the heart of the problem – 'instrumentalist' theoreticians have as a result of increasing experimental data constantly to face up to increasing complexity in the calculation rules they take into consideration. The greater part of their activities consists quite specifically in the attempt to counter this increasing complexity, or more exactly, to counter what they see as its negative aspects such as the profusion of independent parameters and the excessively complicated equations. The way they deal with this problem is to invent new 'rules' which make it possible to eliminate as many independent parameters as possible, and to find simpler and more general equations. These new rules and equations will of course entail certain implications which are susceptible to experimental testing and which the former ones did *not* entail. We should therefore not be surprised to find that even an 'instrumentalist' approach to science leads to a large-scale process of testing that a fallibilist would certainly approve of. It is hardly necessary to emphasise that the fallibilist would approve of it all the more given that the results are often negative; the number of models eliminated by this process over the last twenty years has in fact been very considerable.

Unfortunately, the fact remains that what has been devised by instrumentalists and tested in this way is highly technical. By its very nature, which I have tried to suggest, albeit in a very rudimentary way, this highly technical character more or less excludes all 'thought' in the sense in which 'thinkers' understand the term. Those who maintain that physicists do not think, or no longer do so, thus seem to have some justification for their strictures. The truth here is perhaps that since there is a high level of instrumentalist technical sophistication which science apparently cannot legitimately avoid, there is a gap of some kind between the theoretical physicist's activities and his thinking. Either he thinks or he develops physics. If this is indeed so (which we should have to consider, for the idea is too simplistic not to be an exaggeration) then a real problem has been raised. But – and the point is worth stressing – it is a *different* problem. It is not a matter of sterility. I think I have shown that there is scarcely a trace of such sterility here, and that the fallibilist criticism of instrumentalism in that connection is quite unfounded.

Second criticism: instrumentalism is 'subjectivist'

It is rather surprising to find the charge of subjectivism raised against Bohr, Pauli and Heisenberg by a philosopher of the standing of Karl Popper. According to Popper, these physicists were influenced by the positivist ideas of the young Einstein, and so became subjectivists. Einstein's defection, he goes on to tell us, occurred too late: 'physics had already become the stronghold of subjectivist philosophy it has remained ever since' [3].

A criticism of this kind should not detain us for long if we take it literally. No physicist can take seriously the idea that physics is 'subjective', for the word suggests that some of the data physics provides depend on the subjectivity of this or that particular human operator. In this connection the only legitimately defensible proposition about physics is that it is *intersubjective*. By that term we should understand merely the fact (which I have elsewhere[5] also referred to as 'weak objectivity') that some of the basic statements of contemporary physics do indeed mention the observer, *but on the clear understanding that these statements are deemed to be true for any observer*. The restriction underlined here is obviously an essential one, for it means we can also say that physics aims to present a quite 'objective' synthesis of *communicable human experience*. And an aim of this kind is seen as acceptable, indeed as sufficient, even by many of those who flatly reject the idea that there is no essential difference between scientific research and mere subjective opinion. So if we leave aside the distinction just made and confuse intersubjectivity with mere subjectivity, we make a mistake that is likely to spread confusion throughout the system of ideas. Once again, in view of Popper's considerable acquaintance with the problem, his use of the word 'subjectivist' in this context is rather strange. It seems clear that what he is really talking about is the notion of intersubjectivity.

Admittedly, there remains another unresolved problem: how to justify a kind of intersubjectivity not reducible either to mere subjectivity or to some implicit 'metaphysical realism'. The problem is certainly a real one. But because it is philosophical in nature, involving a matter of principle, it will be reserved for discussion (in substance it is touched on, but alas not solved, in the Addendum).

Third criticism: 'horizontalism'

In the schematic description given above of what could be called 'instrumentalist inventiveness', there was an intentional omission: the use the instrumentalist physicist makes of certain pictorial representations that he knows are naive but which nevertheless do aid the imagination. These are mostly drawn from atomic theory, as was recently the case concerning the notions of 'partons' and quarks. The layman sometimes views such pictures as attempts to describe, within the framework of a realist view of things, the structure of reality as it actually is, quite independent of human experience (to which is attributed the purely passive role of gathering accurate information about parts of that reality). But that is not how such pictures should be viewed. In so far as the physicist imposes consistency on his work as a whole, his introduction of, say, the quark does not mean that he is claiming to construct a theory that, in the sense suggested above, would be any more 'realist' than the more or less 'instrumentalist' theory of the electron already propounded by Bohr, Pauli and Heisenberg. In certain cases, he even confirms, by indirect means, that quarks follow the rules of the very same quantum mechanics that was invented for electrons and which forced physicists such as Bohr to take the instrumentalist conception seriously (if only because of the ambiguities implied by interpreting it in a more realist way). It is nevertheless true that such pictures, even if reduced to the status of what we might call 'parables', are of considerable help to the physicist in generating new ideas. They may for example suggest equations that are more general, or lead to the elimination of parameters – developments which, as we said earlier, provide the true basis, once they have been successfully tested, of progress in present-day physics.

Even if we grant these pictures the rather lowly status described, it must be admitted that they go beyond the framework of a strict methodological instrumentalism limited to rules of calculation. Recognising the important role they play in the process of discovery may lead us on to think that pictures capable of being taken literally might lead to greater progress. So far though, experience has not confirmed that conjecture, and attempts to reformulate physics on a 'realist' basis have produced quite disappointing results over the last sixty years. In the abstract however, there is nothing absurd in persevering with such attempts, especially when they broaden the range of sources of inspiration for ideas. If, as it seems, the atomistic picture has now reached the limits of its intellectual and imaginative fruitfulness, we shall no doubt have to look for others. Those who favour such quests are quite right to stress that if physics keeps too cautiously within the boundaries set by instrumentalism it will run the risk of developing only, so to speak, on one level – that of what can be tested in the short term. This danger of 'horizontalism', such critics maintain, can be avoided only if we are bold enough to formulate theories that are so far beyond the bounds of what is usual as to be *at present* neither verifiable nor refutable.

Like all extreme ideas, this one has its appeal. It may however be rather too attractive, be merely facile beneath its bold colouring. The truth, unfortunately, is that it is remarkably easy to come up with revolutionary ideas in physics – so long as one does not also have to come up with a way of testing them. This is why the objection of horizontalism, although important, needs to be treated with some reservation. We should also note that such an objection applies to fallibilism as well as instrumentalism, since it is the validity of the very criterion of (short-term) testability that it questions.

Remark

As certain authors, notably C. Chevalley [5], have shown, the relations between the physicists of the Copenhagen School who founded quantum mechanics and the positivist philosophers were not as simple as is often thought. Not only did the positivism of the Vienna Circle (nor even, except to a very limited extent, the one of Ernst Mach) have no direct influence on the physicists in question, but they themselves explicitly insisted on maintaining on several major points a clear distinction between their own thinking and that of the positivist philosophers. Thus we find Heisenberg for example claiming in his memoirs that for the positivists 'a ready-made recipe exists; for them the world is divided into what can be said clearly and what cannot be said at all' [4]. But this, he feels, is the most absurd of philosophies, since almost nothing can be said absolutely clearly, and if we eliminate everything that is not absolutely clear we shall probably end up with tautologies totally devoid of interest. In another passage he expresses agreement with the positivists and pragmatists in their insistence on the need for care and precision in matters of detail, and for explicitness in language; but as far as their prohibitions are concerned, he thinks we should take no notice of them: 'for if we had no right to reflect on and talk about the grand correlations we would also lose the compass which enables us to take our bearings'.

Besides, as early as 1927 Bohr and Heisenberg were already implicitly setting themselves apart from what was to become the Vienna Circle. In their report to the Solvay Congress they stressed that reference to observability, an essential feature of quantum mechanics 'does not entail endorsement of the principle that it is possible, even necessary, to draw a sharp distinction between what is observable and what is unobservable'. This is so, they claimed, because once a system of concepts is given one can derive from observations some conclusions about things which strictly speaking are not directly observable; hence the dividing line between what is observable and what is not becomes quite indeterminate. Observations of that kind, of which there is certainly no shortage in Heisenberg's writings, might lead us to think that *initially*, when the conceptual framework of quantum mechanics was being put together, the constant reference to the 'essentially observable' was purely tactical, simply a result of the fact that such a framework was not yet fully constructed.

Thus, as Chevalley notes, various declarations by the theoreticians of the Copenhagen School could incline us to the view that for them reference to observable phenomena 'is neither of neo-positivist nor indeed of purely empirical significance. What it suggests, is neither the application of a principle considered specific to physical theory in general, nor even effecting a return to experimentation' (nor, of course does it suggest a return to experimentation used as a way of countering the ever-increasing importance of mathematisation in physics). In fact, we could even agree with Chevalley's claim that various things said by the physicists in question seem to show that initially they saw 'the methodological reduction of the objects the theory deals with to essentially *observable* physical quantities . . . as . . . a tactical move', as a means, in short, of making it possible to introduce a formalism radically different from one based on intuitively visualisable models.

All this is perfectly true and more than enough to show that it is a mistake to see the *aims* of the founders of quantum mechanics as being the same as those of the positivist philosophers. In Heisenberg's work at least, one can even see the makings of a metaphysics, such that it is not at all absurd to say that *at heart* he was a 'realist', at least in one sense of the term.

Can we go further? Given Heisenberg's desire to construct a new formalism, can we begin to see him as a partisan of mathematical realism? That would be going too far. In the first place, if he ever was a realist this, when all is said and done, was essentially manifested in his efforts to catch a glimpse, on the receding horizons of physics as it were, of what he called 'the grand correlations'. Heisenberg and his school clearly succeeded in establishing a new formalism, but neither they nor their successors managed – as we shall see later in this book – to produce a theory whose fundamental concepts could be considered to correspond to elements of a reality completely independent of human knowledge. Moreover, their sense of the need for such a theory was never so great as to affect their research in any very specific way. They were, it appears, more aware of the fact that such a goal is extremely difficult if not impossible to attain than were the philosophers of their time, or indeed many of those starting out in physics, even today; whereas it was precisely that aim, that feeling for the possibility of an interpretation of a specifically realist kind, that largely determined the direction Einstein was to take in the 1920s in his search for theoretical explanations different from those offered by quantum theory. In this connection, the argument that Bohr and Heisenberg used as the basis for one of their principal theses, that of the completeness of quantum mechanics, is revealing. It is founded on the view that a formal system is complete if it provides the possibility of unambiguous prediction of the results of every conceivable experiment (in a given domain). This clearly means – the point is subtle but essential, and *not* a purely tactical one – that physics is fundamentally grounded in observation. This can plainly be seen in the constant references to the experimental apparatus that form the substance of Bohr's reply to the basic objections made by Einstein, Podolsky and Rosen.

From the point of view of the history of thought, this touches on a particularly delicate question. There is of course no possibility of dealing with it fully in a few lines. Perhaps this is a suitable place to point out that this book is not historical, even less so psychological, in its approach. Observations such as the above are therefore merely parenthetical, aimed solely at avoiding two all too common misunderstandings. One is to see, quite wrongly, the Copenhagen School as holding one and the same view as positivism in its pure and rigorous form. The other is to go to the opposite extreme and unthinkingly imagine that its 'philosophy' is instead totally synonymous with some kind of mathematical realism. With regard to this latter it is essential to give greater weight to what the physicists of the Copenhagen School actually achieved than to the general ideas they put forward at some stage or other of their work.

2
Positivism and fallibilism: philosophical controversies

Most of the criticism that has been made of positivism has come from philosophers and has been directed against philosophical positivism. The positivism of the physicists – the 'philosophy of experience' indeed shares some of the features of philosophical positivism, but in several respects remains quite distinct from it. Any attempt to apply such criticism to it must therefore take into account both the similarities and the differences. For that reason it is not inappropriate to begin this chapter by comparing the main features of the positivism we have been examining with its philosophical counterpart.

Though some commentators trace the origins of philosophical positivism back to the eighteenth century, it was Auguste Comte who gave the word popular currency. It was in Comte's celebrated law of the three stages that this first positivism found its essential formulation. According to Comte, human history first passed through a theological stage, in which all causes were personalised in the form of gods, genies and various spirits. Then came a metaphysical stage, in which attempts to explain phenomena were still made, but this was now done by invoking the existence of *entities* of a very general and often undefinable kind. Finally, a positive stage was reached, characterised by recognition of the futility of asking *why* and a consequent limitation to the pure and simple *description* of facts.

There is no need here to review the influence this kind of positivism had on the development of science in the nineteenth century. It is enough to recall that although much of it was beneficial, some was not. For example, it delayed for some time the use of the notion of atoms, which positivist philosphers in those days saw as belonging too much to the realm of explanation and not enough to that of description. Nor is this the right place to go into the details of the many kinds of positivistic theories of knowledge which emerged at that time (despite the fact that some of them, such as that of Mach, had a very significant influence of research in physics, particularly that of Einstein). But we do need to devote some attention to the positivism of the Vienna Circle, of which mention has already been made. In the interwar years this philosophy, often rather ambitiously put forward by its adherents as *the* philosophy of science, influenced scientific thinking to a great extent. Although this influence has since been vigorously contested, particularly by fallibilists, 'socio-epistemologists' and the like, it is still alive – largely for want of something truly adequate and sufficiently convincing to take its place.

1 The positivism of the Vienna Circle

As initially formulated by its main founders, Schlick, Hahn and Carnap, the positivism of the Vienna Circle is built on one or two very general principles. One of these, declared extremely vigorously by members of the circle, is that the only true knowledge is scientific knowledge. Some people try to reduce this to a mere definition of what we agree to call 'knowledge' (or 'true knowledge'). If we adopt this view of the matter, we must admit that there is nothing at all hypothetical about the principle. However at the same time we must grant that in our culture there *are* some philosophies that assert the existence of realities independent of human beings and to which human beings have access only in non-scientific ways. The proposition that *by definition* knowledge reduces to that which is acquired by means of science cannot logically be identified with the claim that such philosophies are simply false, for that claim is a statement, not a definition. Perhaps the proposition means that we must distinguish between 'knowing' and 'having access to' – but that too would need further explanation. Or perhaps it should be interpreted according to the idea that the verb 'to exist' as used in the philosophies in question has no meaning. But as we shall see later, even that idea runs into very real conceptual difficulties. So on all these grounds it seems reasonable to continue to regard the proposition in question as a true principle – that is to say, as a statement *ex hypothesi* assumed to be correct (or what comes to the same thing, deduced from ideas similarly assumed to be correct).

A second principle of the Vienna Circle positivists is the contention that knowledge must ultimately be capable of complete expression by means of quite precisely defined logical signs and propositions. Adherence to this principle explains their stress on the importance of semantic and syntactic analysis, the latter aimed at eliminating all undecidable or self-contradictory statements, such as those involved in the paradox of the liar. There is no denying that the Vienna Circle saw this as an essential part of their teaching, which is why its members paid so much attention to the problems of language (indeed, this accounts for the name 'logical positivism' by which their philosophy is often designated). Without in any way questioning this historical fact, I should simply like to point out that with regard to the matters discussed in this book, the principle of perfect precision is not of primary importance here.

What *is* here of primary importance, by contrast, is a principle which when all is said and done lies at the very heart of positivism, namely *empiricism*. It can be stated as follows: *sensory experience is the sole source of knowledge*.

This principle needs clarifying, and a way of doing that is to start from the traditional distinction between analytic and synthetic statements. As is familiar, analytic statements are statements whose validity can be appraised solely on the basis of a consideration of the rules governing the language in which they are formulated. By definition they can have no bearing on the content of knowledge. Synthetic statements, on the other hand, do bear on the content of knowledge. To appraise their truth or falsity clearly requires consideration of the source or sources of our knowledge; and of course it is at this point that the principle of empiricism assumes fundamental importance. While for Kant the very general laws of human reason, originating in the structure of our minds rather than in experience, ranked among the genuine sources of knowledge (which implied that in his view there were such things as synthetic *a priori* statements), for the positivists of the Vienna Circle the one and only source of knowledge is experience. Consequently they categorically rejected the idea that there could be any synthetic *a priori* statements. Thus for example the statement that there is one single universal time, which Kant deemed true *a priori*, independently of experience, is for a positivist of the Vienna Circle something quite different, namely a statement of which the truth or falsity must be established empirically.

2 The positivism of the philosophers and the positivism of the physicists

So far the differences we have noted between the two kinds of positivism have been scarcely more than differences of emphasis. If we put aside the axiom made so much of by positivists, that all knowledge can be reduced to scientific knowledge (a maxim about which a number of physicists have some reservations, but which plays no part in the practice of research), then the differences we can see at this stage amount to no more than the fact that the physicists are less radical in their elimination of the synthetic *a priori* than are the philosophers. Contemporary physicists of course also reject the Kantian synthetic *a priori*, deemed to be established once and for all, and to be the immutable precondition of human apprehension of anything; but most of them have no deep-seated objection to retaining the idea that our human understanding has its own schemata. No doubt these have been acquired, but this acquisition occurred over the hundreds of millenia over which our brains developed, and so the schemata in question are impossible to modify in a century or two. Consequently many physicists would hesitate to condemn outright the idea that they are the basis of all possible communication concerning both what people have done and what they have observed when they have done it. In other words, such physicists are receptive to the idea that, as a result of the experiments they conduct, scientists to some degree force nature to express herself in the language of our genes, or at least that they can decipher only those of nature's messages which have a meaning in that language. So, these physicists are receptive to a kind of *de facto* Kantianism, which unlike Kantianism proper is perhaps compatible in principle with the ideas of the positivists of the Vienna Circle, but of which there are few traces in their work.

That may be so. But how, it may still be asked, can this *de facto* Kantianism, this reference to old concepts, be reconciled with the central claim of the positivist physicists, the idea that constructing modern theories also means constructing new concepts and modifying or even abandoning some of the old ones? As it happens, reconciling the two is a simple and perfectly rational matter; one that I would even say would be obvious to all if all had a chance to experience life in some great research centre in fundamental physics. In such places, all that the experimental teams need in order to co-ordinate their handling of massive, complex machines and to keep each other up to date on their results are the concepts derived from our ancestral experience (Euclidean spacc, universal chronometric time, localised objects, and so on), and scientific or technical concepts (energy, current, voltage, and so on), which are derived from and supplement the former without changing them. But right beside such experimentalists are teams of theorists, who talk in terms of curved space, virtual objects, anti-matter, particles not localised in space, and so on; people, in short, who express their ideas with the help of new concepts that seem to contradict the traditional ones. Still, a permanent exchange of view is observed to take place between the two teams of people; they seem both to understand and to need one another. When we see all this going on, it is not hard to appreciate that in order to make sense of the mass of data provided by the experimentalists, the theorists have to create new concepts, that these concepts are unambiguous due to the operational methods used, and that due again to these methods, they are grounded in the set of the traditional concepts which they in some sense extend. In quantum mechanics – and in the many branches of physics based on its principles – this grounding of new concepts in older ones through reference to experiment is particularly striking. Indeed, the notion for example of some intrinsic property of an atom or particle (such as being in a certain place, or having a given velocity relative to some frame of reference) is deemed meaningful only in connection with actual or possible measurements of that property (as we shall see, even the idea of defining it simply by means of a measurement that *could* be made instead of the measurement actually made lies somewhat too far outside the experimental context).

At this point we come close to another difference, in emphasis at least, between the two kinds of positivism. That of the Vienna Circle scarcely seems to question the universal relevance of the concept of a property of a particular object, once that object is seen as belonging to a class of objects for which the property can 'in principle' be defined. Thus Carnap, for one, stresses that the properties of things should be seen as some physical attribute they *have* [6]. It is true that elsewhere, when he introduces the idea of a linguistic framework, he stresses that in his view to adopt what he calls 'the linguistic framework of the world of things' is in fact a matter of free choice. We are free, he says, to go on expressing ourselves in terms of *things*, or to choose instead to express ourselves in terms of sense data. But the point here is that when he says we can choose the first option (and, he goes on to say, we have in fact done so since we were children) he fails to specify that our freedom in this matter is fully operative only with respect to macroscopic objects macroscopically defined. By omitting that qualification Carnap induces us to believe that our freedom in this respect also extends to the microworld and that therefore everything proceeds as if decriptions in terms of things (and of the properties of those things) could, if we so choose, be raised to the status of absolutes. Here, philosophical positivism seems less revolutionary when compared to the old systems of 'near realism' than the positivism of the physicists.[1]

Before going on to a further difference between the two kinds of positivism, we need to be fully aware of one essential point on which they agree. Both see a scientific statement – or more generally a scientific theory – as being meaningful only if it has consequences that can be empirically tested.[2] Further, the ensemble of such consequences *exhausts the meaning* of the statement or theory in question. Thus for example from the point of view of their formulation and of the algorithms they involve, two theories may be very different, even apparently incompatible, but nevertheless be scientifically equivalent if they have the same scientific consequences. (The 'wave' and 'matrix' mechanics of the 1920s are such an example.) The point here is that, in this way of seeing things, a scientific statement or theory which has no experimentally verifiable consequences is neither true nor false, but simply meaningless. What is called 'the principle of verifiability' is just the claim that if a statement or theory is to be meaningful it must be experimentally verifiable.

Of course we still need to define exactly *how* such verification is to be effected. In this connection there are two points to consider. On the one hand, it is necessary to specify in just what circumstances a given consequence deduced from a theory (or a theoretical statement) can be said to be *directly verified* by experiments; indeed, whether such circumstances exist must be investigated. On the other hand, the procedures by which such verifications are carried out must be surveyed critically, and whether, speaking rigorously, it is *possible* to carry them out must be considered.

Much thought has been devoted to these questions; both have aroused intense philosophical controversy. It is not part of the aim of this book to give a history of that debate. Times have changed since then, and now, fortunately, we are in a position to dispense with much of the wavering in the solutions put forward in the past.

Essentially, the first question really asks whether there are statements that can be directly verified empirically, and if so, what are they. Philosophical positivism, we know, gave various answers. Initially the answer was yes. Schlick, Hahn and Carnap adopted what was in fact the common-sense position, which holds that statements about observations are directly verifiable, or in other words that it is possible to know for sure whether they are true or false. Shortly afterwards however, Neurath, also a member of the Vienna Circle, put forward a philosophical counter-argument to the effect that even a statement apparently exclusively concerned with facts inevitably contains an element of theoretical interpretation and that (as Maurice Clavelin [7] puts it quite nicely) the very idea of an interpretation necessarily implies that of a possible *rectification*. Here, it is true, there is a problem of principle, a problem which was to form the basis of one of the two major philosophical arguments the fallibilists brought to bear against the 'justificationist' claims of the Vienna Circle positivists, that is to say, against their thesis that certainty can be attained. Since this critique also applies to the positivism of the physicists, we shall consider at a later stage the conditions under which it is relevant, noting here though that it is not ultimately compelling in character.

The second question – that of the method of verification and the degree of confidence it inspires – can be tackled in two stages. The first consists of eliminating one possible objection relating to the verification of a theory by verifying all its consequences. In substance, this objection consists in asking whether, even if by some miracle we managed to verify empirically with complete certainty *all* the consequences of a theory T, would it not still be possible that there should exist another theory T' with the same consequences? And if so, how could we begin to decide whether T or T' is true?

This is an objection that could indeed legitimately be raised by, for example, those who favour mathematical realism (see above). However, for the instrumentalist (in the strict sense of the term) it has no substance; as we have seen, the very question by means of which it is formulated is not meaningful. In the circumstances under consideration here, both T and T' are true, since the totality of the experimentally verifiable consequences of a theory exhaust its meaning.

The second stage of our analysis of the question at issue here consists in noting that the theory to be established is evidently universal in ambit, precisely because it is a *theory*. We can assert that it is true only if one way or another we are certain that its various experimental consequences are true not only here and now but everywhere and always. However it is clearly impossible to verify today that those consequences will still be true tomorrow. The best we can do is verify them today. This means that even if we admit the possibility of direct verification we still need, if we are to be able to speak of the certainty of a theory or theoretical statement, to postulate that the principle of induction is valid.

The fact that positivists in particular (though not positivists alone) are compelled to fall back on induction in order to establish whether theories are true is what underlies the second major criticisms the fallibilists make of their position. Accordingly we shall return to this point later when we look in more detail at the philosophical criticisms. But let me point out here that it is particularly in connection with answers to the problem of induction that the positions taken by positivist philosophers differ from those taken, implicitly at least, by positivist physicists.

The positions of the philosophers can be seen, at least in a first approximation, as proceeding from 'Baconian' induction, the principle of which is as follows. Having detected the existence of repetitive patterns in the succession of natural phenomena, human beings are led to believe that there are constant relationships in nature, and more precisely, to conjecture that certain *laws* exist. These latter acquire greater and greater respectability as corroborating observations accumulate, so that ultimately – and quite rightly, the Baconian would say – they are seen as universal and certain.

As long ago as the eighteenth century this kind of induction was criticised on grounds we shall have occasion to return to. What should be stressed here, following G. Toraldo di Francia [8], is the point – very important, it seems to me – that both in principle and in terms of its results the process of induction used by physicists, or at least contemporary physicists, is far removed from the Baconian kind.

The simplest way of understanding why this is so is to use two examples. For Baconian induction, the classical example is, of course, that of ravens. After successively examining a very large number of ravens and noting they are all black, the user of this method is supposed to formulate a general law to the effect that all ravens are black. If modern physicists or astronomers proceeded in that way, what would they be led to say? If an astronomer (to take an example from Toraldo di Francia) examined n stars chosen at random and noted that all of them had a particular property P, he would conclude that all the stars in the sky have property P. Similarly, if (assuming it were possible to do so) a physicist examined separately 100 000 atoms of ^{238}U for an hour each and established that during that time not a single one decayed, he would formulate a law to the effect that no atom of ^{238}U decays within an hour. But as all scientists appreciate, no astronomer and no physicist argues that way. What the astronomer *in fact* concludes from his observations is merely a law stating that the likelihood that a randomly chosen star does not have property P is of the order of $1/n$ or less. Similarly, the physicist's conclusion from his observations is a law stating that the likelihood that an atom of ^{238}U decays within an hour is of the order of 1/100 000 or less.

As we can see, there is a significant difference between the 'classical' induction of the positivist philosophers and that of the scientist, or otherwise said, between what the scientist is suspected of doing (by philosophers, who believe he indulges in Baconian induction) and what he is really doing. The conclusion the scientist draws from his findings is clearly a great deal more modest than the one he would draw if he reasoned in Baconian fashion. However this drawback – if drawback it be – is more than made up for by the fact that his conclusion is never false, whereas a Baconian conclusion sometimes is – as it would be in the case of ^{238}U, whose atoms, despite their extremely long average life (of the order of several billion years), are known to be unstable. If a sufficient number of them were observed, some would indeed be seen to decay during the hour in question.

We just saw that with regard to induction the positivism of the physicists is more cautious than that of the philosophers. The final thing to note is that it is also more consistent. This is because the notion of induction is clearly linked closely to that of causality, so that any questioning of causality will bring some discredit on induction. Given these circumstances, the Vienna Circle's rejection of synthetic *a priori* judgements – and hence of causality in so far as it is woven from such judgements – is hard to reconcile with the importance that school assigns, on its own admission, to induction. Since (as we have seen) the positivism of the physicists is *not* associated with rejection of synthetic *a priori* judgements, in all essentials it escapes that particular difficulty.

3 Fallibilist criticism and a critique of that criticism

We have already noted that when all is said and done the positivism of the physicists stands up pretty well to criticism which could be called 'criticism of the first order' since it goes so far as to cast doubt, for various reasons, on the practical functioning of the methodology that goes with it. We have seen that, contrary to the theses propounded by some authors, instrumentalism is neither sterile nor subjectivist (though it is intersubjectivist, a very different matter). We have also seen that ambiguities about the choice of a theory are in fact more apparent than real, given the instrumentalist definition of theory equivalence, and that the commonest objections to Baconian induction are not transposable as they stand to the procedures actually followed by contemporary physicists.

But even if doubts of a practical nature about the trustworthiness of the methods used by instrumentalists can be dispelled, the fact remains that criticism of a more philosophical nature has been levelled against 'positivism in general'. This criticism is currently deemed extremely serious by many philosophers of science, perhaps even conclusive; so we cannot ignore it here. In line with the general plan of this book, we shall consider it exclusively from the point of view of physics. That is to say, we shall *not* take into account the questions of principle that might be raised by the application of instrumentalist methods to other sciences. In addition, to limit further the scope of a subject that would otherwise get out of hand, we shall restrict ourselves to contemporary physics. Within these limits the domain we shall examine has still generated an abundant, rich and constantly evolving literature that cannot be summarised here even briefly. This is why I cannot in this chapter hope to do justice to many important and ingenious developments, but must restrict myself to observations on what seem to me to be the outstandingly important points.

These may be grouped under two headings. One has to do with the notion of observational terms, the other with induction.

The question of observational terms

There is no need here to review in systematic fashion all the objections made to the Vienna Circle positivists by linguists and philosophers, who accused them of having believed, for a time at least, in the existence of terms that are purely observational, and so in the validity of the notion that some basic statements (such as accounts of experience) are *certain*. Once again the literature on this subject is so vast that any such review would be so lengthy as to risk throwing the present work out of balance. However, we do need to appreciate that there are objections of that kind, and that from their own point of view some of them are of value.

Given this, it is a rather surprising feature of such objections that they are often couched in unwarrantedly simplistic and radical terms, which obscure rather than reveal their true import. In this respect let us consider for example the view that our registering of 'raw' sense data and the intellectual constructs we base on them are two quite distinct things. As we shall see, such a view merely expresses somewhat too radically what is essentially a sound intuition. N.R. Hanson nevertheless describes it as 'logical butchery' [9], while Popper in his first major published work, *Die Logik der Forschung* [10], has no qualms about declaring that, far from being certain, the pure and simple descriptions of observations (by which he means scientific observations or experimental findings) are in fact, from a logical point of view, just *arbitrary* decisions.

Such declarations are surprising. Before we discuss them – and more so, of course, before we consider adopting them as the basis for a new way of understanding science – we need to know more clearly exactly what those who made them had in mind. It is therefore worth noting for example, as Radzintsky points out [11], that Popper's position in the English translation of the work mentioned above is already rather more finely shaded: there the claim is made that from a logical point of view the statements in question are accepted by means of a free decision or act. Later (in Schilpp's edition of *The Philosophy of Karl Popper* [12]) Popper explains that these statements are arbitrary decisions to the extent that the accepted judgement cannot be logically derived from any other statement.

We can now see clearly that what Popper wanted to emphasise, in a rather striking way, by this 'outburst' was perhaps ultimately no more than a point of strictness that should be of concern mainly to logicians: there are several reasons (such as the possibility of delusion) why an observational statement is never rigorously derivable from anything that is indisputable. Similarly forceful declarations by various other writers could be interpreted in like manner. While a superficial reading might lead us to suppose that what is being presented to us is a discovery likely to challenge the validity of the very foundations of science (and of the technologies based on it), a closer examination shows that what we are really faced with is in fact a point of logical rigour. The question raised could in fact be compared with the problem of rigour that long concerned mathematicians when they sought to reformulate infinitesimal calculus, or to incorporate Dirac's 'delta function' into their analyses, or again it could be compared to the logician's problem of the 'decidability' of arithmetic. In these and similar cases, the matters in question were clearly significant and important, demanding a great deal of time and effort from those who investigated them. But on the other hand, the physicists who put their trust in infinitesimal calculus before its foundations had been established with all due rigour were quite right to do so, and the same goes with regard to the use of the delta function. And the undecidability brought to light by Gödel's theorem in no way renders the mathematics we use in scientific and technological research any less reliable. The position is more or less the same for the problems associated with observational statements.

The reader looking for a fuller discussion of this point is referred to Appendix I. But, to repeat, it is solely a matter of logical rigour. The point needs study, and this can only be done in any serious way by going into what appears to be rather esoteric detail, as exemplified by the discussions summarised in that Appendix. Such detailed study is quite legitimate, even necessary for professional philosophers of science, whose business, after all, is to look into such problems. Beyond a certain stage though the progress they can make in solving them, however significant it may be, is no longer useful for our purpose here. That purpose, it will be recalled, is to show that instrumentalism is based on, and leads to, assertions about communicable human experience whose validity cannot reasonably be disputed. It seems to me that what I have said above together with what is elaborated in Appendix I suffices to justify the claim that the basic assertions, those on which the method is based (namely, observational statements), are *de facto* beyond doubt. At any rate, I judge this claim to be justified once we accept that intersubjective agreement – for example about raw observations of the pointer of the same instrument, made by several people – may be considered as a real 'first datum', the nature of which need not be examined in any more detail. We need now to move on to a similar justification of the assertions to which scientific research *leads*, rather than those on which it is based.

The problem of induction and the fallibilist approach

It is common knowledge that David Hume was the first to raise what today we call the problem of induction – although he himself scarcely used the term. His own view of the problem was that we use induction constantly in our daily lives, that in many ways we cannot do without it, but that we have never managed to produce arguments solid enough to justify it. Thus, he says, the common-sense view according to which using induction *now*, in this or that set of circumstances, is somehow rationally justified by the fact that *in the past* induction always produced the right results is not an acceptable argument. For that in itself in just another use of induction, so we are drawing on the very principle whose validity we want to establish. And that is an archetypical vicious circle.

That may well be so. But is it really necessary to 'justify' induction? On this question there seems to be some difference of opinion between the majority of scientists on the one hand and the majority of philosophers on the other. Philosophers tend to say that without question it is necessary. That is why textbooks on philosophy – or the conscientious ones at least – describe the various attempts to solve the problem made by Hume himself and later by Kant and others, *together with the critical refutations* of these attempts. More recent attempts, such as the inductive logic of Carnap, Reichenbach and others, have not fared much better. The result is that as things stand at the moment it has to be admitted that the hopes philosophers entertained of a universally-accepted justification of induction remain unfulfilled. On the other hand, the fact is that most scientists are not much bothered about providing a justification for induction in general. As we have seen, they certainly try to keep its use within reasonable limits (and later we shall look in more detail at how they do this), but that is as far as their interest goes in the matter. Rightly or wrongly, their view is that to try to 'justify' everything – including methods as widely used as inductive inference – ultimately results in trying to prove something while denying yourself anything from which to start.

Under the circumstances, it is not surprising that over the last 50 years or so most of the new work on the problem has come from philosophers, and that their critiques have been directed chiefly at other philosophers. Thinkers from all schools have contributed to the debate, but the most incisive criticism of one school of thought by another, criticism that really had an impact, was undoubtedly that brought to bear on the postivitists of the Vienna Circle by the *fallibilists*, or *critical rationalists*.

As with any enterprise worthy of the name, there is both a positive and a negative side to fallibilism. On the one hand it attacks and seeks to demolish its opponents' positions; on the other it establishes and elaborates new notions to replace the old ones. Here the negative side is an extension of Hume's approach to the problem. Fallibilists, following Popper, both posit that we cannot use inductive methods without first offering a rational justification of them, and stress that there is no such justification to be had. This leads them to reject every conception of knowledge that can be described as justificationist. By that term they mean all approaches to knowledge – be they purely 'rationalist' in the restrictive sense of the word, such as those of Plato and Spinoza; or empirico-rationalist, as were the approaches of Galileo and Newton; or purely empiricist in the sense of classical English empiricism – according to which it is possible to build up a body of knowledge that, once acquired, will remain valid for all time. This rejection, which is primarily based on the uncertainties of the inductive method (but also, as mentioned already, on the uncertaintics of the empirical basis), extends to the positivism of the Vienna Circle, since as we saw that philosophy is induction-based and claims to be capable of leading to certainty.

But criticism of a particular approach assumes its full force only when a more satisfying approach is put forward. To decide whether this is the case here, we need to examine the positive side of fallibilism – its declaration that a body of knowledge with a genuinely rational basis can indeed be built up, and further that it is in fact gradually being built up in the form of scientific knowledge, without recourse to induction.

Science, according to Popper, manages without induction. How can this be? Quite simply because, first, contrary to what all systems of rationalist or empirical thought have maintained, it does not, or at least should not, aim at achieving certain, positive knowledge, and second because *negative* certainties – the fact that this or that claim is false – can be obtained from experimental data (assumed in the argument to be certain) without recourse to induction.

This latter point is of course the crucial one. However there is nothing mysterious about it. Quite the contrary: it is obvious – or at least appears to be – as soon as we note that in order to establish with complete certainty a universally-valid claim of the kind 'all ravens are black' it would be necessary to carry out an infinity, or at any rate an exhaustive series, of observations, whereas to refute it one single observation (of a raven that is not black) is quite sufficient.

Fallibilists invoke the essential asymmetry they see between verification and falsification in order to propose an entirely new conception of scientific investigation. In their view, science should abandon the ideal (which in any case is unattainable) of achieving knowledge that is both certain and definitive. Science can however bring about an increase in knowledge, and indeed does gradually do so, by proceeding by means of successive eliminations, that is to say, by formulating several theories and then eliminating some of them through experimental testing. Those theories that survive these tests, or more precisely those that *a priori* were falsifiable by means of some such experiments and yet were *not* falsified when tested, thereby acquire legitimate weight and are said to be (provisionally) corroborated. Thus, fallibilists point out, Aristotelian mechanics was refuted by Galileo as a result of the experiments he carried out, and his mechanics (in its subsequent generalisation as Newtonian mechanics) was later refuted by experiments involving high-speed particles and so had to be replaced by Einsteinian relativity. And so the process continues. There is *no* definitive theory, just as there is no definitive knowledge.

Since our task is not to give a detailed account of the fallibilist position (see for example [13]), I shall merely note one or two characteristics that follow from its essential features. For instance, fallibilists judge a claim or a theory to be scientific (or not) exclusively in so far as it is falsifiable, that is to say, to the extent to which it clearly leads to the prediction of some experimental results which, if they are not subsequently corroborated by experiment, lead to rejection of the claim or theory. Correspondingly, a theory is of (scientific) interest to the extent to which it is productive of such predictions. Thus for fallibilists, scientists should always try to construct extremely bold theories, those which in terms of current thinking are the most 'revolutionary', since these are the ones that *a priori* are most likely to be falsified. The corollary of course is that if they are *not* falsified they will provide the impetus for the greatest progress.

As we have just seen, fallibilism is not a purely theoretical notion. It leads to certain ideas about how research should be conducted in practice. Although questions of method are somewhat outside the scope of the present investigation, it is worth noting in passing what those involved in such research – physicists, that is – tend to think of such ideas.

With regard to the last-mentioned idea, it should be pointed out at once that they have reservations. The recent history of physics would seem to show that Popper's recipe is rather simplistic. It is indeed true that when quantum physics was first constructed extreme audacity paid dividends. But nearer our own time, with the emergence of quantum field theory and of relativistically covariant electrodynamics, a conservative approach proved most productive, since in that case the theoretical framework of quantum axiomatics is retained. It therefore seems clear that faced with the difficult task of choosing the most suitable methods, present-day physicists cannot expect much help from fallibilist recipes of the kind described above.

Nevertheless, virtually all contemporary physicists would accept the idea that if it is to be scientific a claim or theory must above all be *falsifiable*. Once physicists have constructed a theory, the first thing they try to do is derive from it consequences of which it can be said that if in experiments they are shown to be false, the theory will have to be abandoned. They also believe that the greater the number of such consequences the better. However it is also true to say that researchers already thought that way long before the invention of Popperian 'critical rationalism'.

We should now ask ourselves – still from the point of view of the physicist – what the appropriate attitude to fallibilism should be, considering its content rather than its practical implications. The question is of course very complex and cannot, at the present stage of our inquiry, be investigated in all its aspects. In particular, fallibilism has a very important 'ontological' component which we have not yet considered and which we must defer until Chapter 4. Any balanced evaluation of fallibilism must take this component into account. Here, once again, we are obliged to restrict ourselves to two – very important – remarks.

The first has to do with the problem of falsification. Is it really true to say as the fallibilists do that Newtonian mechanics was refuted by experiments carried out on high-speed objects?

As we have already seen in Chapter 1, Heisenberg did not think so. Nor do a great many physicists, who would stress first that there is no theory claiming to formulate claims that are quantitatively correct[3] to every decimal place; for any claim ultimately comes down to statements about the results of measurements and any measurement is flawed by lack of absolute precision. Second, they also point out that every theory has a domain of validity, that is to say a domain, not arbitrarily extendable, of facts to which it applies. Given these two observations, it is clear that Newtonian theory is still perfectly valid within its own domain, essentially that of low speeds (compared to the speed of light). The *only* thing that experiments on high-speed objects show in this connection is that the domain in question does not extend to such speeds.

Mutatis mutandis, such reasoning holds good for most theories, at least as long as no ontological hypothesis is introduced. Once we do that things can easily change. But what we must be absolutely clear about is that as long as no hypothesis of that kind is introduced the existence of a process of successive refutations of theories is by no means an established fact. Admittedly (this is the second of the two remarks) the domain of a new theory includes and goes beyond that of an older one (as relativistic mechanics did with its Newtonian predecessor); the new one generally replaces the old one, at least as far as the physicist's research practices are concerned. This means that it is not wrong to say that *there is no definitive theory*. And when fallibilists content themselves with that particular truth, we must admit that they are right. Given this last remark, we cannot *entirely* reject the support of their ideas that fallibilists claim to find in the history of physics. Nevertheless, it can be said that taken as a whole, the remarks just made do *greatly relativise* it. (Here again this conclusion attains its full weight only if we ignore any ontological concern, by which I mean any attempt to describe anything that could be called 'reality independent of the human mind', or in brief, 'independent reality'. The conceptual landscape certainly changes as soon as we aspire to a *physical realism*. But the question then becomes whether or not that aspiration can ever be achieved.)

That then is my first observation of the content of the fallibilist conception. My second is more technical in nature. Many have written about it, and it is connected with a very penetrating analysis by Pierre Duhem as long ago as 1906. However it needs to be made again here, for it plays a part in later discussion. It is that if we are to be thoroughly rigorous, the basic claims of fallibilism as presented above cannot be sustained. For if we decline to resort to inductive methods we cannot with full certainty refute a claim by experimental means, at least not if, as is generally the case, the means involves use of an instrument. In such cases we cannot be absolutely sure that the instrument worked 'correctly', or to put it another way, in its 'usual' fashion. Clearly, we can be certain that it has only if we rely on induction.[4]

4 Conclusion

From these brief observations and what has been said in Chapter 1, we can, it seems to me, draw some conclusions. As long as we stick to the purely epistemological level and do not introduce the idea of an independent reality which we explicitly seek to describe, we cannot see fallibilism as having a decisive edge over instrumentalism with regard either to the practice of research in physics or to the solidity of its basis. The fact that the overwhelming majority of theoretical physicists continue to use instrumentalism (even if, indeed all the more so, they do so mostly in the way Molière's M. Jourdain used prose – without knowing that he was doing so) argues rather for seeing that methodology as superior. However, instead of regarding the two doctrines as in opposition, it would perhaps be better to highlight those aspects of them that are similar or complementary.

The two doctrines are alike at least in insisting on the experimental foundations of scientific knowledge, and even more alike in seeing scientific research as an eminently rational activity. Neither sees any place for the irrational (or more exactly, the 'a-rational') except at the stage of the more or less spontaneous appearance of 'good ideas'; the processes of verification on the one hand and falsification on the other constitute developments that are entirely under the control of experiment and reason.

The doctrines are complementary in the sense that in practice instrumentalism inculcates in the physicist a necessary modesty about the objectives he should set himself: not *change* as an end in itself, but rather the patient testing of theories somewhat outside their known domain of validity; noting possible discrepancies between what the theories say and what is observed experimentally; searching for theories with a wider domain of validity; and so on[5]. Similarly, instrumentalism readily arouses in the physicist a feeling, in many ways fruitful, for the limits of human powers of knowledge and hence makes him aware that if he does not want to get lost in the clouds he must avoid imprudently building up theories about independent reality that amount to no more than the proverbial house of cards. On the other hand – and it is a question of balance – fallibilism makes up for anything excessive in the modesty that is needed. Its way of highlighting the fact that there are no definitive theories can act as a stimulant for those physicists whose instrumentalist training makes them over-inclined to spend their entire lives in an ascetic pursuit of the elucidation of this or that minute detail.

Finally, to come back to induction, we should note that the fallibilist, together with other philosophers of quite different tendencies, have helped to make plain the deficiencies of Baconian induction. Indeed, although it is true that Hume's arguments apply to induction in general and not only to the Baconian kind, the various examples used – by Popper in particular – are almost all connected with that latter kind. It is by using such examples[6], chosen from among those where there is the risk of error, that Popper seeks to persuade us that even at the level of practice (and not just that of pure logic), inductive methods are frail and unreliable. Clearly, in so far as examples of that kind are convincing it is only on Baconian induction that they can cast doubt. In this connection they need to be examined alongside various paradoxes, such as those of Hempel and Goodman (see for example Toraldo di Francia [8], and Chapter 3 of this book) which have been discovered relatively recently and are to my mind more convincing. There is no gainsaying that their discovery can only strengthen the reservations – the reasons for which we have seen – that many commentators harbour about this kind of induction. This means that ultimately the fallibilist's objections to the positivism of the Vienna Circle – that, implicitly at least, it seems to have recourse to Baconian induction – cannot be dismissed.

However, we have seen above that instrumentalism, the physicist's positivism, does not use that form of induction in its basic procedures and is therefore immune to Popper's arguments against it. It is true that there remains Hume's general objection, that with all due rigour no form of induction can be justified. We have already seen what physicists think about this, namely that basically there is no real need to raise the problem. Most physicists, although they may not have examined their reasons for thinking so, are in total agreement that Hume's objection does not warrant serious questioning of the claims made by instrumentalism.

If we were to bring out into the open the implicit reasoning that leads them to think thus we should find, I think, that it is grounded in three ways. For some physicists, their reasoning is grounded in their opting for a *realist* view of things; later we shall have to consider the possibility of a link between such a view and the justification of induction. For others – there are perhaps more of these – the grounds are those mentioned in Chapter 1. As we saw there, even physicists who are not thoroughgoing Kantians consider that the only thing we can hope to achieve in physics is an interpretation of nature by means of the language which as human beings we inherently possess. In their view, the only means of expression available to physics is precisely that of a language that should be broadly conceived as a deciphering grid. This grid, which may well have been acquired by our ancestors, is at any rate something we cannot do without, and it would therefore be irrational to claim to do so. And – still according to the view of this second group – the related notions of causality and induction are inherent features of that grid. Thus, like Kant, members of the second group see induction – at least in its restricted form of spatio-temporal induction – as an *a priori* form of our reasoning faculty, which consequently does not have to be deduced from anything else.

A third group of physicists adopt the following point of view. They note on the one hand that induction plays an indispensable part in the lives we lead (even animals seem to use it and could not do without it), and on the other that in the sciences spatio-temporal induction at least has never let us down. In a rather Humean way, they suggest that we should take the calculated risk of continuing to use it, adding (implicitly or explicitly) that by convention we should agree to consider the conclusions that step leads us to as 'certain'.

Now such an attitude would never truly satisfy a rigorously logical philosopher. Just as with the problem of observational terms, specialists of that kind are merely doing their job when they try to breathe more logical rigour into the analysis of fundamental problems of this kind. We must of course grant that, but it still seems legitimate to say that, when all the analyses described in this and the preceding chapters are taken into account, the general attitude of physicists to the question at issue does seem to be soundly based. Neither the problem of empirical foundations nor that of induction is of such a kind as to raise serious doubts about the validity of instrumentalist methods.[7]

3
Border areas of instrumentalism

What we have seen so far has progressively led us to believe that instrumentalism as practised by physicists is in fact a more solidly-based methodology than many would allow, and that this conception – or method – does indeed stand up very well to the objections of an epistemological nature that have been made against it. This does not, of course, imply that it represents ultimate truth; in fact, in the second part of this book we shall, for very good reasons, try to go beyond it. But before examining these further considerations, which are fundamental to our purpose, it is appropriate that we should develop somewhat further our ideas about instrumentalism by looking at one or two areas in which, though the conception remains coherent and highly acceptable, some of its limits are discernible.

What follows is therefore a transitional chapter, one that though not vital for an understanding of subsequent material nevertheless enables us better to appreciate certain points and clarify particular details. We shall be concerned with the notions of properties and laws, as well as – once again – the idea of induction, regarding which the earlier promised clarifications will be produced.

1 Properties and laws in instrumentalist physics

Within the philosophy of experience, or to use the expression adopted in the preceding chapters, of 'instrumentalism', the notions of *properties of physical objects* and *laws* take on meanings which do not exactly match those given to them in 'common-sense' realism. This is inevitable. In effect, 'common sense' assumes we can deal with reality directly, that we can therefore define laws merely as regular patterns which govern the relations between the components of reality, and regard properties as being possessed by these elements of the real. The instrumentalist conception sees the facts of the problem rather differently. If it is thorough-going, instrumentalism gives no precise meaning to the expression 'independent reality' and so cannot, it follows as a matter of course, define laws and the properties of things by referring to reality. Even a merely methodological instrumentalist finds himself in this position, for he knows he can make no statement about reality until the concepts he would have to use for that purpose have been discovered. In particular, this is the position of the physicist working at the level of atomic and subatomic particles. Given this, we need to be aware of some at least of the positions taken up by instrumentalists (in the widest sense of the term) with regard to the notions of properties and laws.

Most of what follows in this section has already been covered in my earlier book *In Search of Reality* [2]. Those parts of it which may be useful in understanding later parts of the present book are reproduced here.

The notion of properties

Concerning this notion the general requirement that experience should be the ultimate reference clearly means that we have to introduce an operational definition. This is so even in the simplest cases. To say for example that a given object is red, is, by definition, to posit that if the object is illuminated the observer is aware of redness (we can ignore details such as the use of light containing visible radiation and so, as they are irrelevant to the argument). Even there though a question arises: if the object we are discussing is for the moment not illuminated, can we still say it is red? This question may well remind us of the old philosophical debate about what can be said concerning the relationship between colours and objects, but this is not the place to approach it in that particular form. What is significant from our point of view is that the question is a very general one, and that it appears in the writings of modern philosophers when they deal with 'dispositional' terms as well as in those of physicists seeking to define the notion of property at the microscopic level.

The problem of dispositional terms has been well treated by Hempel [14]. A dispositional term is any term defined as a disposition of an object to react in such and such a way or to produce this or that effect in particular circumstances. Hempel chooses as his example the term 'magnetic'. If as above we leave aside details irrelevant to the argument, we can say that the meaning of the adjective 'magnetic' applied to an object x is operationally defined by the assertion that '*if* iron filings are placed in proximity to x, *then* they are attracted to x'.

Following Carnap, Hempel points out however that for anyone concerned with linguistic precision and anxious to eliminate ambiguity, a definition of that kind has an undeniable flaw. This has to do with the precise meaning of the phrase 'if X then Y', where X and Y are two propositions. Formal logic – which must be our guide, logical positivism tells us, when we seek to get rid of the ambiguities of language – gives that phrase a rigorously defined meaning, that of *material implication*: either X is not true, or Y is true, or both. In other words, in terms of the logic we are discussing the assertion made (that is: if iron filings are near x they are attracted to it) is true in two sharply distinct cases. The first is when both X and Y are true, that is, when the filings are there and are attracted. The second case is when X is false, whatever may be the case as regards Y.

Hence, in particular, the assertion is true whenever there are no iron filings. Consequently, one of the implications that formal logic says should be drawn from the assertion is that any object x around which there are no iron filings is *ipso facto* magnetic. Clearly this is not at all what we have in mind when we say that something is 'magnetic'. It follows that ultimately this particular assertion cannot be used in its present form to define the dispositional term in question.

It is obvious that here we are faced with a general problem concerning the mode of defining any dispositional term. To get round it, Carnap proposed a very different mode, which he called 'the method of partial definitions'. A partial definition of 'magnetic' is provided by, for example, the assertion that 'if iron filings are in proximity to x, then x is magnetic if and only if they are attracted to it'.

From the point of view of formal logic, such a definition does not raise the difficulty noted above. On the other hand, it is obviously only 'partial', since it defines the meaning of 'magnetic' as applied to x only in the particular case of iron filings in proximity to x. To establish the totality of cases in which we say that an object is magnetic, Carnap and Hempel suggest giving several partial definitions of but one term. Thus to the partial definition already given, we would add for example the statement that 'if x passes through a loop of copper wire, then x is magnetic if and only if a current appears in the loop in question'. We could continue in this way until we had taken into consideration all the qualitatively different situations in which we say that an object is magnetic.

What does the physicist think of such a procedure? What, more generally, does he think of the problem that lies behind it? The answers to such questions are likely to differ somewhat according to whether they come from a macroscopic or a quantum physicist, especially if the latter has addressed the problem of the foundations of the kind of physics he works on. His 'macroscopic' colleague will very likely answer without hesitation that for him the problem does not arise. From this point of view, the difficulty Carnap raises is simply due to the fact that Carnap resolutely sticks to an over-restrictive definition (the so-called 'material' implication) of the expression *if . . . then . . . .* It is not therefore a real problem, or at least is one which in so far as it is real is of interest only to philosophers of a certain stripe, but is quite irrelevant to the actual practice of science.

The response of the quantum physicist, however, is likely to be rather less clear-cut. Indeed, the more he has studied the positions taken up by earlier colleagues of the Copenhagen School, in particular that adopted by Niels Bohr in his reply to Einstein, Podolsky and Rosen on the problem of measuring observable phenomena, the more likely he is to be favourable to Carnap's theses. The fact is, as shown elsewhere (ISR [2], Chapter 5), that the restriction to partial definitions Carnap deems necessary in connection with dispositional terms, and the restriction to precisely-specified experimental apparatus Bohr finds indispensable in attempts to give precise form to the notion of properties (as elements of reality) in connection with a microscopic system, are closely akin. Generally speaking, talking about the properties of such a system 'in itself' (that is, independently of the experimental apparatus) ultimately has very little meaning for Bohr, who seems to see such properties, with certain exceptions, as having no existence in scientific terms, and to feel that mentioning them in a scientific context can be conducive to erroneous reasoning.

The fact that the leading figure of the Copenhagen School and the positivist philosophers thought along similar lines may seem rather surprising. It has already been stressed that while there are some convergences, these are mainly differences between the ideals of the positivism of the philosophers on the one hand and the instrumentalism (at least in the methodological sense) of twentieth-century physicists on the other. Those (and here again I include myself among their number) who do not find the position of the Vienna School satisfactory can in particular legitimately point to the fact that the kinship we are talking about here does not constitute an identity. The physicist's position on the question of attributing properties to quantum objects is too finely shaded for us to reduce it to a systematic and universal application of the method of partial definitions advocated by Carnap and by Hempel. The point will have to be discussed further. On the other hand, those whom positivism makes uneasy might also be tempted to claim that Bohr's position, now half a century old, is out of date. That charge though is true only to a very limited extent. Of course it *is* true that in a field such as relativistic quantum formalism there has been remarkable progress over the period in question. It is undeniable, although to a lesser extent, that there have been significant advances in the theory of the measurement of observables. Nevertheless – and in part *because* of these very developments – the intuitions of Bohr, Rosenfeld and other physicists of the Copenhagen School concerning a certain indivisibility at the conceptual level of the quantum system and of the instruments which measure it are now seen to have been largely justified. In cases in which it would seem natural to speak of the precise value a of a physical quantity A pertaining to a quantum system x, present-day physicists are often led, in accord with Bohr's perspective on the matter, to decline to do so unless x is associated with a clearly-defined measuring device which is capable of measuring A. More precisely, they are often led to assert that in other situations it is meaningless to talk of a; this is particularly the case with regard to the (unknown of course) value that *would* be found if a measurement of A were made, though it will not *actually* be made.

It is a pity that it is not possible here and now to go into a detailed explanation of the factual reasons that make theoretical physicists feel obliged even today to stick to positions so close to what we might call the Copenhagen spirit. These reasons are rather complicated and technical;

this book is not the place to deal with them in all the detail an expert would demand. However we shall outline them after our examination of realism in Part II. What we need to note here, in order to clarify some things, is simply that in the early days of quantum mechanics the 'intuitions' noted emerged in connection with the celebrated wave/body duality. As is now familiar, any microscopic 'particle' has both wave and corpuscular aspects. More precisely, the 'particle' appears sometimes as a body and sometimes as a wave. In other words, it is appropriate to speak of it either in terms of bodies or of waves, *depending on the experimental apparatus being used to study it*. It was this methodological observation that the Copenhagen School was to raise to the status of a principle, that of complementarity. The idea that to attribute properties to quantum objects independently of the measuring devices is ultimately meaningless can be seen as one facet of this principle of complementarity.

Counterfactuals and laws

As just pointed out in connection with the notion of properties, present-day physicists see themselves increasingly obliged to go along with those philosophers who are determined to restrict themselves to the method of partial definitions, the notable feature of which is to avoid definitions expressed in terms of conditional assertions. We can express this more exactly by saying that the method refuses to have recourse to counterfactual implications. Whereas 'ordinary' formal logic recognises only *material* implications (the 'if . . . then . . .' referred to above), there are more elaborate kinds of logic (which in a way are closer to the actual mechanisms of thought) that involve other types of implication. In particular, the *modal* kinds of formal logic introduce the notion of *strict* implication, derived from the notion of *necessity*. Taking necessity into consideration is what gives these so-called modal logics their specific identity.

These variants of formal logic are based on the idea that there is a distinction to be drawn between the mere affirmation of the truth of an assertion (which may be purely contingent, as in Valéry's classical example 'the Countess went out at 5 o'clock') and the assertion that a proposition is necessarily true (as in the example 'five is an odd number' or 'this proton has a spin ½'). There are, as we can see from the two last examples, at least two kinds of necessity: the 'logical' and the 'physical'. Other kinds could be thought of, but here we shall pay particular attention to *physical* necessity (that is to say, to implications that are strict only in so far as the laws of physics are taken into consideration). In modal logic the term 'strict implication' refers to any (material) implication which is in fact necessary. 'If John is a bachelor, then he is not married' is a strict implication of a logical order; 'if an iron bar is heated it will expand' is a strict implication of a physical order.

If an implication is strict, that is to say, if circumstances *A* necessarily – inevitably – entail event *B*, then there are certain consequences. Indeed, even if situation *A* does not actually obtain, the mind can always imagine it does. Hence, knowing that the implication is strict, the mind is led to see the statement 'if *A did* obtain, *B would* occur' as meaningful. In the present hypothesis (that *A* in fact does not obtain), the statement is an instance of *counterfactual implication*. Consequently we can see that a counterfactual statement is essentially a kind of strict implication. (Bearing in mind the very precise definitions that logicians are obliged to give to the terms involved, this is true in all rigour only if certain precise detailed conditions apply. There is no point setting these out here; see for example [16].)

The definitions of the various *properties* of physical objects, which we spontaneously and for the most part implicitly formulate, are practically all based on counterfactual implications. When we say that a pebble is hard, we mean that *if* we tried in one way or another to split it into several parts, we *would* find it difficult to do so. When we say that sugar is soluble, we mean that *if* we left it in water for some time, it *would* dissolve, and so on. It is very easy to verify that replacing material implication by a counterfactual implication enables us to avoid the paradox described above. And as we shall see, we then have in principle a mode of defining properties that cannot but appeal to realists.

Given the problems referred to it was interesting to note as we did that in the first half of the present century epistemologists by and large managed to eliminate all counterfactuality, even all modal logic, from the process of defining properties, simply by using partial definitions. (Their success was not complete; it is still difficult to define certain terms, such as 'soluble' and 'adaptable', in that way.) But it is noteworthy that in the circumstances they did not attempt, or more accurately did not consider it legitimate, to perform the same feat with regard to the nature of the definition of physical laws. They (Carnap [15] in particular) devoted a great deal of attention to the concept of law and were above all concerned to stress the difference between physical laws and what they call 'accidental universals'. In their view, this difference is very closely related to the idea of conditional implication. What, they asked, is the difference between a statement such as 'on 1 January 1958 everyone in Los Angeles was wearing a purple tie' (an accidental universal), and statements of the kind 'when heated any iron bar will expand' (a physical law)? Their response was not straightforward but one difference they did stress is that a physical law makes it possible to justify a counterfactual claim (in this case, that the iron bar in question, or any other, *would* expand *if* heated) whereas accidental universals do not (here the universal in question would in no way make it legitimate to assert that if I had been in Los Angeles on 1 January 1958 I would have worn a purple tie).

It is worth drawing attention to such references to counterfactuality – and hence to modal logic and the notion of causal necessity – in the writings of people such as Carnap, who otherwise affected scepticism about precisely that kind of necessity. Perhaps this development in their thinking can with some justice be seen as an example – one among many – of the recognition that there is a certain complexity in the relation between the world and the mind apprehending it. The desire for perfect clarity of language and the rejection of vague ideas (one of which, in the view of many adherents of logical positivism, was that of necessity), unfortunately put up artificial limits to the exercise of thought. On the other hand, such an observation of course does not mean that we can simply do the exact opposite of what is counselled by such a prudent attitude, and uncritically endorse ill-defined intuitive notions. We shall have the opportunity to assess the dangers of such an approach later, in our study of realism.

2 Induction and confirmation*

Compare the following two statements. The first is fairly short, the second rather more complex.

Statement 1 If it is true that each time conditions *C* are observed to obtain, phenomenon *P* is observed, then the law 'conditions *C* imply phenomenon *P*' is increasingly confirmed as the number of cases *N* in which conditions *C* obtain and phenomenon *P* is actually observed increases.

Statement 2 Let s and s' be two space-time regions that are 'congruent', that is, are superimposable by means of space-time translation and spatial rotation, and let *C* be the set of initial and boundary conditions which in s are accompanied by the occurrence of phenomenon *P*. It can then be stated that if in s' the same conditions *C* all obtain, phenomenon *P* occurs also.

We know from Hume's argument that it is not possible to provide a rigorous justification of induction, but it is clear that we can nevertheless contemplate linking the principle involved in it to a statement of one of the two kinds given above. Hence there is some point in approaching the problem of induction by asking which of the two seems to us more appropriate when we are simply trying to provide a fairly general basis for the method of inductive inference.

At first glance it might seem that there is no doubt about the answer: that from this point of view statement 1 is much to be preferred over statement 2, which, however useful it may be in everyday life, is not capable of general application. Quantum phenomena violate it – and it is worth noting in this connection that saying this is not merely a question of interpretation. Even if we imagine a larger set of initial and boundary

* This rather technical section can easily be left to a second reading.

conditions by hypothesising the existence of extra variables ('hidden' parameters), this will not remove the violation, as we shall see in Chapter 5, because of non-separability. The fact of the matter is that statement 2 would hold good with all due rigour only within the framework of a physics at once deterministic and 'local'. But the only known physical theories (and one could, it seems, almost say: the only conceivable ones) to which those two qualifiers can be attached are theories subscribing to physical realism. Many people would therefore consider that statement 2 is plausible only within the framework of physical realism. On the other hand, statement 1, for its part, is quite general in nature. *A priori* what it says seems to be so blindingly obvious that it is hard to imagine its content giving rise to any kind of argument.

But it does. Far from being invulnerable, statement 1 is open to criticism pertinent enough for many to have stopped using it for any purpose, and in particular do not use it as a starting-point in their analysis of induction. In Chapter 2 we took the example of the uranium atom to illustrate one of these criticisms. Another arises from the existence of a paradox that is implicit in the statement, as Hempel has shown. It lies in the fact that saying that every time C is observed P occurs is logically the same as saying that every time that P does *not* occur C is not present. Now according to statement 1, the second claim is further confirmed every time it is observed both that P does not occur and that C is not present. By virtue of the logical equivalence just pointed out, this confirmation also applies to the statement in its original wording. So if we consider that the physical law stating that all unsupported stones fall is confirmed each time a released stone is seen to fall, it seems that we must also admit the same law is further confirmed each time a motionless stone is seen to be bound to some support, which is absurd.

Spatio-temporal induction

For reasons we have just noted, scientists such as the physicist Toraldo di Francia [8] maintain that we should abandon statement 1 altogether. However, their account of the concept of induction is not directly based on statement 2, which as we have said is too particular. Following them to some extent, we shall therefore try to construct a new principle of induction by, as described in [8], suitably generalising statement 2, in particular by changing it into one that is also applicable to indeterministic physics.

To do this, it is natural to think of treating probabilities as if they were physical properties like any other. Since in physics (all due respect being paid to mathematical rigour) probabilities finally appear as limits of frequency, we are obliged to consider instead of the single space-time region s talked of so far the whole set of space-time regions congruent with

s. Let E be a subset of that set. Let us suppose that a certain set C of initial and boundary conditions are met within each element of E and that, given this proviso, E may be composed in any way. Let f be the proportion in E of regions in which P occurs. Let s', as above, also be some region congruent with s (it may or may not belong to E). The principle of induction we are seeking can then be expressed as follows: whatever region s' may be, if conditions C are met in s' the likelihood p of P occurring in s' is roughly equal to f. The significance of this assertion is that if we decide to examine a number m of s' regions thus defined, we shall find that P occurs in a number of these regions that is approximately equal to $f.m$.

Adopting Toraldo di Francia's terminology, we shall describe the induction thus introduced as 'spatio-temporal'. By comparison with Baconian induction, to which statement I naturally leads, spatio-temporal induction (which of course is much more restrictive in its domain of application) has at least two major advantages. In the first place, it does not lead to Hempel's paradox. In the second, once we have given an appropriately precise meaning – as we should, of course – to expressions like 'approximately' and 'roughly equal to' in the wording set out above, such induction in fact coincides with the quantitative methods of evaluating probabilities, mean lives and so on used by physicists in our own times.[1] It is applicable to the quantum case and also to the classical one when the initial and boundary conditions are known only incompletely. We should note that it is in no way based on the notion of confirmation. *One single experiment* (including of course a great many elementary measurements) on a beam of particles of a certain type enables us to evaluate to a specifiable degree of accuracy the mean life of particles of that type. Confirmation merely serves to reduce the risk of human error.

The details regulating the application of the few guiding ideas summarised here of course bring to light some questions of logical rigour. It is not necessary to review them here. However there is one point we should perhaps consider as an example of such questions. This point, not always appreciated, is that in the statement setting out the rules of generalised induction there should be no confusion about what 'composed in any way' means. It is not a matter of letting us make a totally free choice of the elements composing E, for if that were the case we could exercise our preference for some particular space-time regions differing from others with respect to certain initial or boundary conditions not included in the set C. This would mean we would be introducing into the composition of the sample E a 'bias' that would render the rule not valid in general. We must also have similar reservations about how we should understand the clause 'whatever region s' may be'. In both cases we must suppose that the choice of regions is purely random with respect to conditions not included in the set C.

Discussion

The foregoing analysis seems to lead to a fairly clear conclusion. Statement 1, to which Baconian induction is linked and which seemed to give the latter the ultimate support of 'obvious' truth, does not in fact stand up to scrutiny. By contrast, statement 2 (or more correctly, statement 2 in the generalised form just described) turns out to yield a correct description of 'useful' scientific induction. But many regard it as justified only within a realist conception of things. Should we see this as supporting the realist thesis? My own view is that we should, but that opinion is not held universally. Some specialists refrain from considering statement 2 as a cautious reformulation of the realist-determinist hypothesis, and of course they are equally resistant to considering its generalised form as reformulation of the hypothesis of a statistical determinism based like strict determinism on a realist conception of things. For them the initial and boundary conditions described in statement 2 cannot be *all* the conditions 'really' holding in s independently of human observational capacities, since to their mind such a notion would smack of metaphysics. All that such conditions *can* be, they feel, are conditions that can in fact be observed. In particular, they feel that such conditions can be defined only with a certain margin of error, within which the law has been verified. So according to them statement 2 tells us that, under conditions identical to those in s within such margins of error, the phenomenon we shall observe in s' will be P, to within some other margin of error.†

To those taking such a view it seems that two questions can be asked. The first is: what exactly should we understand by a statement such as 'within a certain margin of error the phenomenon we shall observe will be P'? If this is taken to mean that the phenomenon will always be largely similar to P, statement 2 cannot be true. In view of the uncertainty concerning initial and boundary conditions, the phenomenon observed in s' could well be quite different from P; it is easy to give examples of cases where that is in fact what happens (Frauenhofer lines in the solar spectrum, fine resonances in particle scattering, and the like). It might perhaps be possible to construct another sense of the statement, based on the notion of probability for example, but in that case it would seem that we are no longer dealing with statement 2 but rather with its generalisation, as described in the last paragraph in some detail. This would mean that statement 2 itself would lose its meaning and, deprived of its starting point, the 'generalisation' at issue would seem very abstract. . . .

The second question is none other that the old one raised by Hume: what justifies statement 2 (or its generalisation)? More precisely, what

† This interpretation has been described to me in a private communication from G. Toraldo di Francia.

makes it *non-arbitrary*? Would we say that it is our experience of an enduring similarity between the patterns of various sequences of observations? But how can we say whether this will continue to be the case? When we assume that what happens today will happen tomorrow are we not laying down an *arbitrary postulate*? And if we assume that things will happen in this way the day after tomorrow are we not laying down another postulate, just as arbitrary as the first and quite independent of it? Faced with the series of arbitrary postulates which a categorical refusal of any implicit realist metaphysics would impose on us, we may – quite rightly in my view – prefer the equally arbitrary but single postulate of 'the uniformity of nature'. But if we want to escape from the situation described above and at the same time avoid the evasions that ambiguity too easily promotes, we have to be clear that we cannot take 'nature' to be merely synonymous with 'the ensemble of observations'. In the present context, we may only take the traditional word 'nature' to mean an underlying reality, which may or may not be knowable (that question is here not at issue) but which is seen as existing quite independently of the observer.[2]

If my analysis is correct, it admittedly does not lead to the conclusion that realism truly *justifies* induction. It is well known that, with all due rigour, there is a vicious circle here. Nevertheless, it somehow backs up an idea that is suggested by common sense, namely the idea that it is *because* we think the Earth exists quite independently of our mental faculties, and *because* we have been taught that its motion is governed by definite laws that are also independent of them that we firmly believe it will once more be daylight round here tomorrow.

Part II

Physical realism and contemporary physics

4
Physical realism and fallibilism

Throughout Part I the philosophy of experience (or methodological instrumentalism, or the positivism of the physicists) was defended against the excesses of a critique which on the one hand tends rather bluntly to identify that standpoint with logical positivism, and on the other hand deliberately ignores the solid and wide-ranging advances it has enabled science to achieve. Here I shall not rehearse the arguments already presented, merely stressing my view that they are substantial and deal adequately with rather ill-conceived objections. In no way do I subscribe to the essentials of positivist philosophy. For that very reason however, I deem it vital that only valid and genuinely relevant arguments be ranged against it. In the same way, it also seems to me important to distinguish between instrumentalist *methodology* and positivist *philosophy*. Defending a sound thesis with shaky arguments – and arguments which confuse these two can only be shaky – always leads to obfuscation.

I The inadequacies of positivism

That said, let me repeat that I am not attracted to pure positivism, and here, without embarking on a discussion of the history of philosophy, let me first of all at least *indicate* what seems to me the unsatisfactory features of any instrumentalism that is not linked, even if only implicitly, to some kind of realism.

Reservations raised by instrumentalism

Some of the features of instrumentalism that arouse justified reservations have already been indicated. For example in Chapter I we saw the disadvantages of the high degree of technical sophistication demanded by the instrumentalist attitude. Unfortunately, such sophistication is indispensable; but the scientist who employs it tends, unless he is very careful, to regulate his thinking after a peculiar fashion that is fruitful chiefly (and in certain cases *only*) within his professional field. We also noted in that chapter the well-known danger of 'horizontalism'. Always to insist as a matter of principle that a theory must have consequences that are experimentally testable in the very near future (and using existing procedures) may mean running the risk of ossification. Is it not the case that genuinely new ideas, at the moment when they take shape in the human mind, may well have no consequences that can be verified or falsified by technical means? It is plausible that to make major conceptual leaps and establish real connections calls for a less rigid methodology.

Running alongside these criticisms which we have already noted are certain observations that *a priori* might also lead us to have reservations about the wisdom of choosing a purely instrumentalist approach. One of these is that if we do adopt such an approach it becomes impossible to understand *why* a theory enables us to predict experimental results accurately. As Radzintsky [11] notes, there is only one possible answer to the question. A theory enables us to make accurate predictions because it offers a sufficiently accurate picture of certain aspects of reality; because, that is to say, it contains a hard core of truth. Of course, within the framework of a simple *methodological* instrumentalism an objection of that kind is not really an objection at all. We are free to conceive of a more or less veiled reality (a notion to which we shall return) whose 'temporal' regularities, albeit deformed and transposed, are reflected in the laws governing phenomena – thus accounting for the predictive power of those laws. However with regard to instrumentalism laid down as a *principle*, the objection remains relevant (and so is analysed in greater detail in Chapter 9).

Other more specific objections to instrumentalism can be listed. One is that the notion of intersubjectivity (or 'weak objectivity'), the very foundation stone of the instrumentalist edifice, remains obscure. This is because although we have direct, immediate knowledge of – at very least – our own impressions, our knowledge of the impressions of others is only indirectly mediated. There is no obvious logical reason for granting these two kinds of knowledge the same status. This means, as Einstein observed, that the die-hard positivist who claims to follow his doctrine to its extreme conclusions may well fall into the trap of solipsism.

Another criticism of the same kind concerns our knowledge of past events. If it is strictly true that the experimentally-testable consequences of a theory exhaust its meaning, it becomes hard to see what meaning could be ascribed to evolutionary theory. Or rather, we find ourselves more or less obliged to restrict its scope to predicting what we shall be able to observe. What the Big Bang theory says, for example, would only amount to the claim that if we look for an intersidereal radiation of 4 K we shall find it – or to other claims bearing as this one does on experiments in the present. Seeing in the theory any statement bearing on what 'really' happened would be 'metaphysics' and so, in the view of many scientists, 'illusory'. (In this connection, see later chapters, in particular Chapter 12 note 1). Similar remarks may be made concerning Darwin's theory of evolution, and so on. It may be, after all, that such theories should be considered as mere metaphors. But before deciding in favor of such an idea it is certainly appropriate to wait for it to be backed by less *a priori* and less general arguments.

Ambiguities of operationalism

In Chapter 2 mention was made of the criticism of induction fallibilists have deployed against philosophical positivism and more generally against all theses the fallibilists describe as 'justificatory'. I have sought to answer these criticisms already, at least in some measure. However there is one *a priori* reservation which is akin to these criticisms but which very directly concerns the philosophy of experience as employed by the physicist. It bears on the notion and use of *operational definitions*.

As we have seen, the systematic use of operational definitions is essential in quantum physics, and more generally in all circumstances where the uncritical use of familiar concepts is safe only when describing experimental apparatus and cannot be extended without very strict controls to the objects themselves being studied. But there are numerous fundamental problems associated with operational definitions. One of these, already studied by philosophers, has to do with the fact that measuring a physical quantity – length for example – involves several distinct procedures, each of which is specific to a clearly determined order of magnitude. Astronomical distances are not measured by the methods used in evaluating terrestrial distances, nor by those used for molecular distances. From a strictly operationalist point of view, our usual way of regarding such diverse methods as defining a single physical quantity can seem quite unjustified, even if the procedures in question overlap to some degree; it opens the gates to the woolliness of metaphysics. More specifically, in some areas errors can arise as a direct consequence of interpreting data in terms of familiar everyday operations in cases in which the way the data are measured has nothing at all to do with procedures used in everyday life.

This difficulty is clearly akin to the one we have already noted concerning the definition of dispositional terms. In fact, even a preliminary study of the problem (as in Chapter 3) clearly shows that the very notion of operational definition itself can be understood in two quite different ways and therefore needs to be defined more precisely.

This already holds good for the terms relating to macroscopic properties (which are the ones philosophers usually examine). Indeed, as we have seen, a term such as 'magnetic' can be defined either by implicit reference to counterfactual implication, or (and to the mind of the thorough-going positivist, more rigorously) by means of a set of partial definitions. But it is mainly in the field of microphysics that the distinction is important. In the sphere of our usual, or more or less usual experience of things, the field of both macroscopic physicists and philosophers, the difference between the two ways of understanding an operational definition is purely theoretical. Which one we choose has no practical consequences. In the sphere of everyday experience, choosing between them never facilitates or complicates the interpretation of theory or the analysis of experience. The same is not true in quantum physics, as a detailed study of the Bohr/Einstein controversy clearly shows (see for example ISR [2] Chapters 7 and 12). To be sure, it is also common in quantum physics to define the notion of the value of a physical quantity in an 'operational but counterfactual' way, that is to say (as is the case in classical physics) by reference to measurements that could be, but in reality are not actually made. This is what is done when, for example, a particle accelerated in a synchrotron is said thereafter to *have* a definite momentum. This is commonly said even about particles whose momentum we have no intention of measuring. It is then tacitly admitted that the property in question is quite independent of the observational apparatus, and it is considered that what we say is meaningful just because we know we *would*, in this case, get one defined result *if* we measured the property. However, this way of defining properties of microsystems turns out to lead to serious conceptual difficulties, which we shall set out and discuss in detail in Chapter 10. In view of this and also the fact that such questions cannot be settled entirely by experiment or by means of calculations, and hence are epistemological rather than physical, many physicists tend more or less to adopt Bohr's position (already referred to) in considering that this counterfactual method is still somewhat too much of an extrapolation of what we are used to doing in macroscopic physics. They are willing to resort to it for pragmatic reasons if doing so does not raise any particular problem, but will reject it if it does raise problems, using instead an operational means of definition which takes into account the observational apparatus actually used. To put it another way, they then resort to a means of definition which, as we have seen, is very similar to Carnap's method of partial definitions and which we might expect to give rise to the same or indeed more fundamental reservations. Should we, for example, extend the method to cover all objects? Should we, in a very general way, see the very properties of things (and not only our knowledge of such properties) as depending on the observational procedures *actually* used by the human observer? Once again, this *may* be the case, but before we accept a thesis of that kind we should wait and see whether it is supported by any convincing arguments to the effect of ruling out less anthropocentric interpretations of the phenomena involved. And if we do not accept it unreservedly, where should we draw the boundary between the domain of application of the two kinds of definition? Positivism gives us no clear guidance here.

The reservations so far listed about positivism in all its forms are not yet exhaustive. There is a final one, which has to do with the little word 'fact', as it appears in the writings of philosophers. In spite of the enormous lengths to which the positivists will go to avoid any trace of ambiguity, it must be acknowledged that their use of this word does sometimes indeed seem rather . . . ambiguous. This is because in the writings of many of them, a ‘fact’ is sometimes but a short-hand term for an ensemble of perceptions (ultimately shared by the collectivity of human beings), while at other times it designates (in an uncritical way) ‘something’ that belongs to reality; to a ‘reality’ they seem to conceive of as fully external to human beings and, as such, entirely independent of human observational faculties (though nevertheless, of course, still perceptible by these faculties). To be fair, it is quite understandable that philosophers with no more than a general knowledge of subatomic physics should not be aware of the inappropriateness of using the term interchangeably in such ways. For as long as such knowledge remains superficial, Carnap’s thesis that the ‘linguistic framework of the world of things’ (see above, Chapter 2 p.41) can be chosen freely and without restriction will seem tenable. And its consequence, if it were true, would be that we could interpret our experience of microscopic phenomena just as freely in terms of objects as we do our experience of macroscopic phenomena in terms of such objects. In both cases it would merely be a matter of convention. However, a less superficial grasp of the nature of quantum physics provides, as we have seen, much more information to take into account. More specifically, it shows that it is impossible to extend indefinitely the linguistic framework of the world of things into the sphere of the very small. The paradoxical consequence is that even in the writings of positivists, whose prime concern is linguistic exactness and precision, there is often an essential ambiguity in these matters.

Although different in nature and scope, all these reservations have at least one thing in common: they prompt us to make ourselves acquainted with conceptions that, to varying degrees, part company with what we have described as hard-line positivism. These conceptions include ‘methodological instrumentalism’, where the qualification ‘methodological’ means that the conception has room for the idea of a reality independent of what is human. They also include a more specific form of realism, what I call *physical realism*, which lays it down as a principle that the reality at issue can be known unambiguously. Of course it is obvious that realism is a concept that allows of a great deal of fine shading. We shall examine some of its various forms, moving from the most specific to the most general.

A word of warning is in order, however. In proceeding thus, we shall use the technique of *reductio ad absurdum*. The guiding principle of our method will be to consider first of all a rather specific kind of realism, suppose as a working hypothesis that it is true, examine the consequences of that supposition, and then discover that one (at least) of these consequences contradicts known facts. Once the very specific kinds of realism have thus been eliminated, we shall move on to examine others that are less specific in their conceptual foundations. Using this general method, we shall be able to determine the level of generality at which realism ceases to generate difficulties when confronted with factual knowledge.

That at least is the logic of the programme we shall follow. In practice we shall be obliged to cut short a few stages and accept on trust some of the conclusions, either to spare the reader technical details that have no place here (further elucidation of such matters can be found in [2A]) or because they have already been discussed in detail elsewhere (see ISR [2]).

2 Fallibilism and physical realism

Does every kind of philosophical realism have to be derived from 'ontological' principles? Some realists say no. Indeed, many philosophers are well aware just how hazardous a business it is to claim to say anything about 'Being in itself'. So aware, in fact, that even those – or at least some of them – who claim to be realists take good care not to base their thinking entirely on a *realist metaphysics*. Whether in the event they avoid the hazards or whether they are to some extent merely playing with words is not our main concern here. What is beyond doubt are their efforts to found their conception on non-arbitrary notions. Deciding what these notions should be is the problem.

A sensible way of doing this, which in outline seems to have been the method adopted by Popper as well as by many physicists, is to take into account the *degree of corroboration* of major theories, or in other words, the number and quality of tests to which the theory in question has been submitted successfully. As that author said [3], when we are obliged to act on the basis of some theory, the rational way of proceeding is to use that theory, if there is one, which so far has stood up most successfully to all attempts to disprove it. In such circumstances, what better way of proceeding could a realist follow than to adopt the working hypothesis that this particular theory comes closest to expressing the truth, that is, provides us with the best account of the texture of reality?

Provided such an account is not regarded as totally and definitively established once for all, a fallibilist realist can indeed adopt such a procedure. But one condition remains to be met. The theory has to be capable of being formulated in – or at least translated into – strongly objective terms. By this I mean that its basic principles must all be such that they can be stated without any reference to the community of human observers, in particular without any reference to the limits of the abilities of these observers. If this condition were *not* met, the basic principles of the theory in question could not refer to reality. In any given epoch therefore, a person who subscribes to physical realism (such as a fallibilist realist) will, provisionally at least, adopt as the best available account of reality that universal theory which, of all those that meet the above condition, has been most fully corroborated. It is virtually self-evident that in doing so he will be led to consider (always, of course, subject to the possibility of revision) the fundamental concepts of the theory as being consonant with reality. There have been few historical epochs in which two competing theories resting on different fundamental concepts have been equally well corroborated. It is therefore likely that there will be few cases in which this method will not enable us to determine which concepts are to be regarded as best adapted to representing reality.

By and large, the method just described is the one adopted by the great majority of realist *physicists*. Clearly it is rather different from the one that realist *philosophers* have generally adopted since the beginnings of our intellectual history, since such philosophers mostly preferred to present major ideas in axiomatic form, seeking thereby to deduce consequences and to arrive at certainty. Or else, still motivated by the notion of certainty, they thought they had found it fully and completely in the reduction of knowledge to facts – not realising that facts have to be sorted by the mind and, as pointed out above, cannot all be interpreted in a strongly objective way. By contrast the above procedure is, to repeat, common to both the realist physicist and the Popperian philosopher, who consequently is quite close in spirit to the tradition of physics. That tradition, it should be stressed, was always based on a 'pedestrian' approach, which demands that the starting point should be composed of simple, even fragmentary data, capable of being studied in a precise fashion, and that on the basis of these data we should try to formulate successive generalisations by means of cross-comparison, testing and – frequently – rejection. Correlatively, the tradition has almost always incorporated a high degree of objectivisation of concepts, a process very closely matching the objectivisation or ordinary concepts human beings have always spontaneously inclined to. A concept that 'works' – because it is an element of a universal theory that itself works – is tacitly and almost inevitably raised to the status of an absolute, that is to say, of an 'element of reality', by the overwhelming majority of the scientific community. Such is physical realism. It is worth repeating that such a procedure is quite reasonable. Even if we can see its limits, we are more or less bound to grant it our approval.

Nevertheless, we must remain aware of these limits. At the stage we have now reached, there is fortunately no difficulty in doing so. They arise from the fact that theoretical knowledge increases, a phenomenon which often brings in its wake enormous changes in basic concepts. A notion that once seemed fundamental and therefore lent itself to objectivisation (the gravitational force for example) is suddenly reduced to merely an approximate way of speaking. A realist who has adopted the ways of thinking described above must therefore expect to encounter major upheavals in the realm of concepts. *A fortiori*, he must always be ready to engage in successive refinements of the way he sees reality. And the experience of past ages shows that increasingly this process of refinement requires him to put aside his most familiar idea and hence to accept a realism that is 'remote', a realism based on concepts that seem strange and perhaps even unacceptable to the great mass of the uninitiated.

This, it will be said, is the very essence of fallibilism . . . As indeed it is. And so we should take note of something that apparently not every fallibilist has seen: fallibilism in fact *proceeds* from physical realism as just defined, or to put it another way, *it is opting for physical realism that* justifies the fallibilist conception, and not *vice versa*. If we now bring together these observations and the matters we considered in Part I, what emerges? In the first place, we see that the history of physics in no way forces an instrumentalist *who rejects or sets aside realism* to become a fallibilist. This is because *he* sees the essence of science, and of physics in particular, as consisting of experimental data on the one hand and equations on the other, both being strictly approximate and the latter being grouped in formalisms, each formalism having its own, large or small, domain of validity. Now, neither as regards the experimental data nor as regards the equations did any 'breakdown' ever take place in physics. Even the Ptolemaic system is still valid in its limited domain, and in more recent areas equations such as Maxwell's are not only valid but *indispensable* in physics, despite the fact that there is no such thing as an 'ether'. What actually occurs in the history of physics (still speaking from the instrumentalist point of view) is therefore not a sequence of paradigm collapses but rather a steady advance, consisting of successive replacements of a formalism, valid in domain of validity D, by a new formalism which does not necessarily make its predecessor obsolete but which has a domain of validity D' which includes, and is therefore more extensive than, D.

From this perspective, all concepts are relative and their objectivisation at a particular time is ultimately of sociological rather than strictly scientific interest. Consequently, their down-grading to merely metaphorical status in no way means that a theory has collapsed. So fallibilism is irrelevant here. On the contrary, within realism as depicted above concepts are implicitly or overtly objectivised so that their down-grading to the status of mere metaphors can easily be seen as revolutionary and, indeed, as representing the collapse of theory itself. Once we appreciate this and bear in mind that historically events such as those alluded to here have occurred fairly frequently, fallibilism does indeed seem to be an essential complement of the kind of realism we have been discussing.

In section I above (p.68) we noted the ultimately unsatisfactory nature of any instrumentalism not associated with some kind of realism. If the only conceivable kind of realism was what we have referred to as 'physical' realism, there would be weighty arguments in favour of fallibilism. In this book however I hope to make progressively more plausible the case for a different kind of realism, one that is less closely linked, it is true, to the scientific knowledge extant at a particular point in time, but also (and consequently) less likely to be affected by the events we have been considering. But we must not look too far ahead. For such an approach finally to be seen as acceptable, it must first be preceded by a detailed and painstaking examination of the explicit and structured 'physical realism' underlying the works of recent physicists and philosophers.

3 Counterfactuality and physical realism

The kind of realism we are discussing is essentially, albeit somewhat implicitly, linked to the notion of counterfactual implication. More precisely, we could say that it has close connections with the method of defining the properties of systems based on counterfactual implication. As we have seen, this realism is a kind of sublimation of common-sense realism that does not radically change its nature. When in everyday life we say that an object, or an attribute of an object, is real (and, *a fortiori*, when we say it is 'material'), what we are implying is not only that we or others will actually observe it but, more basically, that we *could* observe it, or observe its existence, *if* certain conditions were fulfilled, conditions that may actually not be so. Even if it is not actually seen, a mountain is said to be real as soon as it is granted that if we went to the country where it is supposed to be we would see it. The same holds good for properties. And we conceive of and even implicitly define real events essentially as changes of properties.

One consequence of this is that using operational definitions is not the exclusive feature of strict instrumentalism. Operational definitions based on counterfactual implication (see above, p.59) are in fact more relevant to a realist view of things than to strict instrumentalism. Conversely, if we *accept* that counterfactual implication is a legitimate mental operation, we can, even within a systematically instrumentalist perspective, use it to define new properties and also to specify (as also pointed out above) what is meant by 'physical law'. This means that it would be improper to say, for example, that to a thorough-going instrumentalist the very concept of law cannot be defined. It is nonetheless true that the ideas of necessity, and hence of law, on the one hand and of counterfactual implication on the other seem much less arbitrary and therefore much more acceptable within the framework of a realist, as opposed to a thorough-going instrumentalist vision of things. In other words, there is, in practice at least, a very close link between counterfactual implication and realism.

In fact there is more to be said. Objectivising the *properties* of systems seems to require that the counterfactual implication which the realist uses to define them should cover a relatively large and not precisely defined *sphere of accessibility*.

The notion of a sphere of accessibility is fundamental to both modal logic and the theory of counterfactual propositions. In view of the important part it plays in the question we are investigating here, it is necessary to state what is meant by it, even at the cost of a fairly lengthy digression. To do this, we should first examine in some detail the idea of necessity.

We can all, of course, intuitively grasp the qualitative difference between a true but contingent proposition such as, to once again use Valéry's classic example, 'the countess went out at five o'clock', and one which is *necessarily* true, such as 'seven plus five make twelve', or even 'granite does not burn'. Quite rightly however, some logicians (but not all: there have been those who even denied the basis of the distinction) have tried to state the difference precisely.

To begin with, they noted that the difference in question arises partly at least from the fact that the propositions we describe as necessary are seen as true in all kinds of imagined situations. A novelist telling us that a countess went out at 5 o'clock has in mind a single event taking place on a specific day, and his report does not claim to tell us anything about what she might have done if it had been snowing on that particular day. Quite the opposite, of course, is true of the mathematician who states that seven plus five make twelve, for he is making a statement purporting to be valid for every day, whatever the state of the weather. Statements of the second kind could even be held to be true in 'worlds' different from our own. A phrase like 'different worlds' is more striking than 'various situations' and is therefore rather more appropriate, since the importance of the idea of necessity needs highlighting. That is probably why specialists in modal logic prefer it. We shall use their terms and describe a proposition as *necessarily* true if we hold it to be true not only in our own 'actually existing' world but also in a whole range (composed of an indefinite number of elements) of imagined 'possible' worlds.

The way this array of possible worlds is composed needs more attention. It is at this stage that the differences between various sorts of necessity – logical, physical, temporal, and so on – are taken into consideration. If for example we are trying to define logical necessity, the relevant set of worlds will be that in which the laws of logic are the same as in our own. In the case of physical necessity, it will be that set (subsumed in the former) in which all physical laws are the same as in our world, and so on.

The set of possible worlds we are considering must in any case contain the actual world. It is therefore useful to represent this set symbolically by a solid ball or sphere, with the centre i being the actual world, the radius of which varies in length according to the degree of generality of the worlds to be considered. These are said to be 'accessible' (from our world). Each is represented symbolically by a point within the 'sphere of accessibility' [16]. By definition, a proposition is *necessarily* true in world i (the actual world)

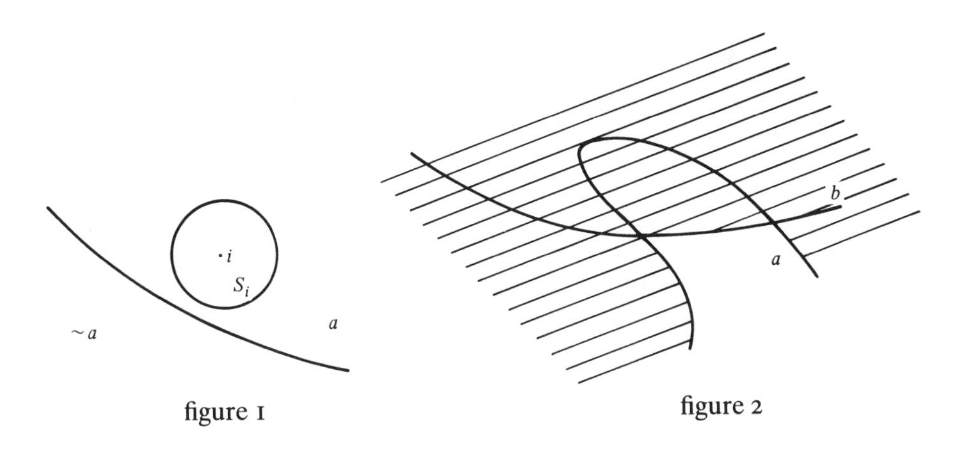

figure 1 figure 2

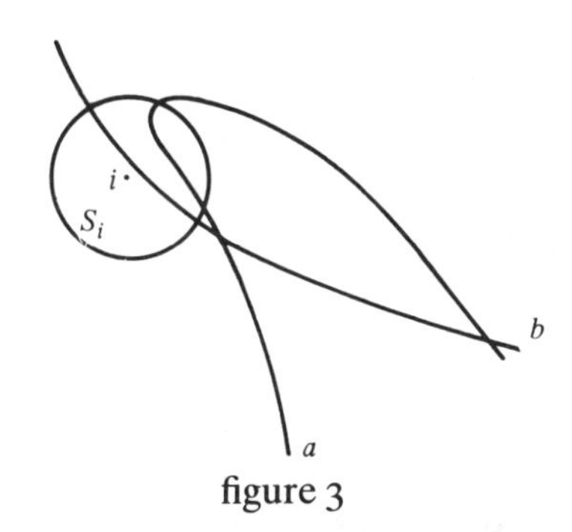

figure 3

if it is true in the whole sphere of accessibility centred on i. Henceforth we shall call this sphere S_i. To show that a proposition a is necessarily true, we shall express it thus:

$\Box\, a$,

thus implying that $\Box\, a$ is true if and only if a is true throughout the sphere S_i (cf. figure 1).

From the ideas of necessity and of

$a \supset b$

(material implication: if a then b) it is easy to define strict implication:

$\Box\,(a \supset b)$,

which reads 'it is *necessarily* true that if a then b', or in other words, 'if a then necessarily b'. Figure 2 expresses symbolically the proposition $a \supset b$, since, as we know, $a \supset b$ is identical to $(\sim a) \cup b$, in which $\sim a$ indicates 'not a'. The shaded areas are those in which the proposition in question is true. If we then refer to figure 3, we note that $a \supset b$ is true in the entire sphere of accessibility S_i, and that consequently the figure represents a case in which $\Box\,(a \supset b)$ is true.

It will be noted with regard to figure 3 that the proposition $\Box\,(a \supset b)$ may very well be true even in cases in which, in the actual world i, neither a nor b is true. This is of course quite normal, since it is quite possible that at the moment a given bar of iron is neither being heated nor, consequently, expanding; this does not falsify the proposition that if it *were* heated it inevitably would expand. In such cases the strict implication is in fact *counterfactual* or 'counter to the facts'; that is to say, is an implication with a premise that is false in the actual world i.

Now that these matters have been clarified, we have reached the end of digression. First, it is clear that a property of a system can be operationally defined by reference to a counterfactual implication. Such is the case for example, with *position* as a property of a particle. Here the sense of the statement 'at time t particle P is within the interval $(x, \triangle x)$ of the axis O_x' will be defined by the counterfactual implication 'if (proposition a) an appropriate instrument for measuring position *were* in place, then (proposition b) it *would* record the presence of P within $(x, \triangle x)$'. Or in symbolic terms,

$$a > b,$$

where $>$ designates the counterfactual implication ($a > b$ can be identified with $\Box\,(a \supset b)$ if certain reservations about details are ignored). But here the precise definition of the sphere of accessibility S_i relative to this counterfactual implication raises difficulties and can give rise to ambiguities. We know, for example, that in the actual world, for reasons of very general validity, *it is impossible* to subject *simultaneously* a particle P to the effects of an instrument suitable for precisely measuring its position, and to one of those suitable for precisely measuring its momentum (Heisenberg's relationships). Should we or should we not then include certain imagined worlds in which P would be subjected to the effects of an instrument for measuring momentum in the sphere S_i under consideration here? Those who would argue against doing so would base their objection on the impossibility we have just noted and which clearly refers to human instrumentation and hence to human beings. Their case would be that in imagined worlds the same impossibility should be assumed to obtain, and that consequently it would be inconsistent to imagine a world in which a device for measuring momentum operates and in which proposition a (the measurement of position) could be true. It is clear that this is a fundamentally *instrumentalist* attitude. Those who would argue in favour of inclusion, that is to say, those with a *realist* turn of mind, would argue that it is quite possible that Heisenberg's relationships refer only to purely *human* limitations, which have to do only with *our* possibilities of knowing the real world, and that when we are trying to define the very *notion* of the 'position of an object' such contingencies should not be taken into account. Some would even add to this that the major interest of the method of defining by means of counterfactual implication lies in that it shows us the necessity of using the concept of *imagined* worlds, and would declare (strong objectivity) that subjecting such worlds to the all too human contingencies of *our* cognitive faculties would mean abandoning the major strength of the method. At any rate, the attitude of those in favour of inclusion, if applied as in our example to the property of 'position', means that this concept will be meaningful in the most general circumstances. A particle (or at least its centre of mass) always has a position. That is the essence of a conception we shall call *microrealism*.

4 Realism and metaphysics

The fact that both the attitudes described above can be defended by rational arguments suggests that seeking to eliminate either solely on the grounds of considerations concerning the part played by counterfactual implication in defining the meaning of words would be a delicate undertaking. Such considerations are useful however, in that they indicate that those with a realist turn of mind tend to think implicitly, even if they are not always aware of doing so, in terms of a much larger – since it is less limited by anything human – sphere of accessibility than would spontaneously occur to those with a bent for 'positivism'. If the universe has a 'negative curvature' (as astrophysics currently appears to suggest), realists do not hesitate to conclude that it is infinite, which clearly implies that it is meaningful to evoke the idea of celestial objects beyond the cosmological horizon. It is true that we cannot conceive of any instrument capable of showing us such objects, and hence imagined worlds containing such instruments are worlds to which we cannot, for fundamental reasons, attribute anything like 'operating' existence. For that reason, positivists will hesitate before taking them into consideration. However, realist physicists (and, curiously, in this field we are almost all realists) are not put off by such a difficulty. This is because what is at stake is not whether it is possible to *operate* (in one way or another), or to observe (this or that fact), but rather whether it is possible to *ascribe a meaning*. Since this is also what is at stake in the problem bearing on the idea of the position of particles, it is hard to see what *a priori* justification there could be for physical realists not to talk, in all circumstances, of the precise position of this or that particle.

On the basis of these two examples, it becomes possible to claim that the method of definition by counterfactual implication, once it has been fully and properly explained, reveals a paradox well known to philosophers. The paradox, or dilemma if that term is preferred, can be expressed as follows. On the one hand, the human mind does not enjoy absolute primacy (it is almost inconceivable that it should, for if it did why not simply adopt solipsism?); but on the other it is absurd to employ words to which we cannot give a meaning. Yet to give a meaning we need to refer to experience, which can only be human. Willy-nilly therefore, we must refer our concepts to human beings, and hence ultimately centre on human beings all the descriptions we try to provide of whatever it is we see as 'real'. The way in which the human mind, in so far as it opts for physical realism, tends to resolve this conceptual dilemma is, as we have just seen, to define the meanings of words via definitions based on counterfactual implications and to give such implications a very large sphere of accessibility with limits unaffected by our human faculties. (We might note here that a significant feature of this explanation of physical realism is that it clearly shows that it is *legitimate* to distinguish between 'having a meaning' and 'being observable in actual fact', a general reflection which will prove useful later.)

It is well known that the second half of the twentieth century has seen a revival in the fortunes of realism. A number of philosophers – in the United States as well as in Europe – have rebelled against the kind of discourse centred on language alone that was offered by their predecessors. More or less citing Popper as their authority, they demanded that things, not only words, should be dealt with. Even if the foregoing developments constitute but one way of interpreting a certain kind of realism[1], it is nevertheless true that, insofar as they appear as justified, they may be used in order to support this revival. They show that *a priori* it is neither absurd nor illegitimate to transgress what we might call the *prohibitions* decreed by positivism, and that we can do so yet still respect the demands of intellectual rigour which must – or at least should – now preside over the way we work out a philosophy worthy of the name. In view of the positivists' claim that any conceivable *physics* should be contained within their edicts, it is not even out of place to express what may be original in the arguments put forward above using a kind of Popperian language, that is, by saying that contrary to the theses for which positivists have argued most ardently, these arguments restore the acceptability of a certain kind of *metaphysics*.

If I might look ahead a little, I should like to say that (as will become apparent in the conclusion of this book) on this last point I agree with what emerges from Popper's reflections. I too believe that some element of what the positivists rejected as metaphysics is essential if knowledge is to be coherent. On the other hand, with regard to what could be described as the *details* of Popper's realism, my position is more critical. Again, it seems to me that what any physicist can approve of unreservedly is *the way Popper proceeds*. He refuses to start from pre-established dogmas that could easily turn out to be mere prejudices, and never assumes that any *description of the real* (even, and perhaps especially, if it assumes the guise of a scientific theory) is certain and definitive. What he does do is pay the greatest attention to the degree to which science corroborates the available descriptions of reality while remaining willing to abandon those descriptions the moment advances in scientific research cease to corroborate them. As I have said, such a procedure, which begins from facts, from the concrete and the partial, and moves by trial and error, fits and starts, towards the general, seems to me to be the very procedure adopted by scientific thought since its earliest days.

But, of course – again this is the essential feature of the procedure in question – it is necessary that when we adopt fallibilism we should accept the possibility of having to give up some ideas we have, perhaps even our most cherished ones. As we have seen elsewhere, it was natural in the early nineteenth century to see the notions of Euclidean space, universal time, forces acting at a distance, and point masses not just as elements of an algorithm (the Newtonian theory) offering high powers of prediction, but as parts of an exact description of reality in itself. That point of view can no longer be defended. More generally, we should note that the gradualist way of proceeding endorsed here shifts our image of the world in the direction of increased generality but also further from what appeared to be basic and obvious facts. It cannot be stressed too much that, despite the hopes of well-known realists, this inevitably entails a parallel evolution of our system of concepts. In this connection it is not irrelevant to point out that the weight such realists give to the degree to which theories (and hence their underlying concepts) are corroborated, and above all the fact that their realism forbids them from assigning human limits to the spheres of accessibility that serve to define certain concepts, both tend to produce what may sometimes be a somewhat sterile conservatism. The first factor is clear. The second has the same effect because of the fact that such an absence of 'human limits' entails, as we have seen, the idea of extrapolating possible measurements into areas where they are in fact impossible. This in turn gives rise to a prejudice in favour of keeping old concepts, those defined by these measurements, even when working out theories intended to deal with such new areas.

These dangers should be avoided. But a clear-sighted follower of Popper who strives to do so cannot rule out the possibility that as a result of successive refutations of increasingly ambitious theories, and hence of increasingly unfamiliar concepts, we may reach the extreme stage in which the human mind can *no longer* construct, in the form of a description of reality, *any* conceptually coherent and unambiguous realistic interpretation of the *very general intersubjective rules of prediction* that it is still capable of formulating. (By 'conceptually coherent and realistic' I mean an interpretation built up on the bases described above, or in short, using only concepts defined in terms of strong objectivity.) This is at least a possibility that cannot logically be excluded.

Now it so happens that the rules of prediction in quantum mechanics when examined in detail are, as we shall see, so difficult to interpret on a strongly-objective basis that some of those chiefly responsible for their discovery immediately let it be understood that the possibility just mentioned might indeed reflect the actual state of affairs. As has been shown elsewhere – and we shall return to this point in later chapters – more recent experimental and theoretical developments have made the possibility even more likely. Microrealism has but a small chance of survival, and those of physical realism are not much greater.

5
Microrealism and non-separability

Our programme, it will be recalled, is to start with very naive kinds of realism and seek to refute them where possible by demonstrating that one or other of the consequences they entail is false. We then take up a more sophisticated kind of realism and repeat the operation on it, and so on. It must be stressed however that in following this procedure, in which stages that are obvious are presented in summary form, great prudence must be exercised. A particular hypothesis may seem at first sight to lead to consequences that are invalidated by observation, when in fact a more detailed analysis shows that this is far from being the case. In the last chapter for example, we encountered some motivations for considering the conception we called microrealism, according to which entities such as electrons, quarks and the like, to which the name 'particles' is ascribed, are deemed to have a specific position at any given moment (and in terms of this conception, should also have, 'for reasons of symmetry', a specific velocity). It might be thought that the existence of the so-called 'uncertainty principle' would in itself be enough to refute this kind of microrealism. Nowadays however, virtually everyone is aware that this is not the case. The 'uncertainty principle' could be *but* a matter of uncertainty.

The fact is that there are several kinds of microrealism, not all of which are equally attractive. Those that at first sight are most attractive are those that obey the 'principle of separability' (to be defined later), but these, as we shall see, have indeed been refuted. Not, however, on the basis of the uncertainty principle, and not on the basis of the existence of interference phenomena either; for refutations relying solely on those considerations can themselves be refuted with no great difficulty.

1 Necessary precautions

To illustrate this last observation, and thereby to draw attention to a kind of erroneous reasoning that is all too common, I propose to introduce (provisionally) an idea that has been abundantly corroborated by *classical* relativistic physics. I have in mind the notion of (physical) separability 'in the wider sense of the term', seen as synonymous with the thesis that reality is composed of entities (particles or field strengths) which are localisable in distinct regions and whose influences on each other decrease as the distance between them increases. (In passing it should be noted that the principle of separability we shall examine later is merely a more precise specification of this idea.) In the view of certain people who know something of the procedures used in quantum mechanics, this kind of separability has already been refuted by the very fact that there is such a thing as a wave function, which is not localised at a particular point and can sometimes extend over great distances. They point out that with Young's slits (an experiment well known from physics textbooks) the wave function passes through two slits at once, and they claim that this 'proves' non-separability. Now when reduced to this level, the argument is in fact worthless. Indeed, in a similar way a single wave in the sea can simultaneously travel around both sides of a rock . . . and this common observation in no way refutes separability 'in the wider sense of the term', since a wave is made up of parts, each of which go round only one side of the rock. The analogy clearly shows the implicit hypothesis which vitiates the pseudo-refutation: unless we can prove that a wave function is not made up of parts, we cannot establish by the line of reasoning pursued here that separability is false.

It is of course extremely difficult to construct such a proof. This has led to the idea of a somewhat more elaborate refutation procedure involving systems composed of at least two particles, which have interacted in the past but at the time in question are observable at points distant from each other (or as classical physics would say, have a great distance between them). When two particles P_1 and P_2 have interacted, the wave function of the P_1 P_2 system in general cannot be expressed by combining *one* wave function relating to P_1 with *one* relating to P_2 (this is an inevitable consequence of quantum mechanics). In fact, an expression for the wave function in question emerges in which *several* functions of the P_1 coordinates (each of which could play the role of a wave function for P_1) are present and similarly *several* functions of the P_2 coordinates (for each of which the same applies)[1]. A consequence of this is that neither P_1 nor P_2 has any well-defined wave function. Only the P_1 P_2 system has one. This is what is called the non-separability of the wave function. Such a structure is responsible for the statistical correlations observed in such cases between the results of measurements involving P_1 and P_2.

Given the essential part played by the wave function (or its mathematical substitutes known as 'kets') in quantum mechanics, there is no doubt that it is legitimate to see the (mathematical) non-separability of the wave function as constituting a highly plausible argument against separability. Nevertheless, however plausible it may be, once again it is *no more than* plausible. *A priori*, at least two ways of circumventing it can be thought of. The first is quite simply no longer to insist that the state of a particle must always be capable of description in terms of a wave function (or ket) and to accept certain other descriptive algorithms, called statistical operators by physicists and of which only one particular type corresponds to wave functions, as equally valid. However this method does no more than shift the problem of separability elsewhere. Indeed, if we use it we have to distinguish between two kinds of ensemble (real and improper mixtures), the existence of physical differences between which I established long ago [17]; and once the distinction has been properly made, something similar to the non-separability of the wave function re-appears. (Further details can be found in either [2B] or [2A].) The second way, however, is somewhat harder to argue against. It is based on the observation that there is no *a priori* obligation to consider the wave function of the system as necessarily constituting a complete description of it. And if we abandon the implicit hypothesis that it does, the above-described argument against separability loses its power to convince; for then the wave function is simply an agent indicating the *proportion* of composite P_1 P_2 systems having certain properties within an ensemble of such composite systems. Under such conditions the correlations observed between the particles in question can be *qualitatively* explained as in the corresponding classical case, namely by reference to common causes.[2]

As explained in detail elsewhere (see ISR [2]), Einstein's claim that quantum mechanics is not a complete description of reality was essentially based on an observation of that kind. The same idea, or more precisely the thesis (called microrealism above) according to which the extra parameters differentiating between particles having the same wave function are the position and velocity of each of them, was, it would seem, upheld for a long time by some eminent contemporary philosophers of a realist turn of mind. Let it be stressed once again first that *qualitatively* such an idea is made plausible by certain epistemological arguments whose nature has been described above, and second that it is such as to enable us (if correct) to retain the idea of a physical separability of systems (the idea we called 'separability in the wider sense' above).

This simple example shows that in this area we must beware of coming to over-hasty conclusions and that we must not let our ideas be guided passively by the formal structure of the mathematical entities involved in the theory. All that certain proofs of non-existence – and more particularly some celebrated alleged proofs that the extra parameters just mentioned do not exist – actually establish, as some clear-sighted physicists have stressed, is that those who produced them were nowhere near as gifted with imagination as they were with the capacity for abstract thought . . . .

But this is not enough to allow us to infer that the principle of separability holds good. In fact this is not the case, as we shall see later once we have properly considered all the data to do with the problem.

2 The notion of an objective state

In the last chapter we looked rather carefully at the realistic hypothesis (namely, that it is meaningful to say that there is 'something' fully independent of our knowledge, it is then that 'something' that we call *reality*). Having noted the extreme generality of this hypothesis (for 'reality' could be matter, God, Ideas or whatever) we quickly moved on to a more specific version of it, namely physical realism, which maintains that reality so defined is scientifically knowable[3]. This version, which includes microrealism, must now be explicated and clarified before we go on to examine, here and in Chapter 6, whether or not it is compatible with experience. To do this we must first define the notion of an *objective state* as generally as possible.

With that end in view, let us posit, merely as a working hypothesis with consequences to be examined, that reality can be considered as made up of distinct but potentially interacting parts called physical systems, or more briefly, just *systems*, and that the individual reality of each of these can be specified by a set of parameters having the nature of real numbers. This set of parameters, which we shall specify by the single symbol λ, specifies at any given moment, and therefore defines, what we shall call the objective state of the system[4]. It should be clearly noted that no specific hypothesis about the nature of the parameters λ is being proposed, which means that our 'working hypothesis' is compatible with many theories. In the older physics – that is, when the world was seen as reducible to a set of point masses interacting by means of either contact forces or forces decreasing with distance – the λ would have been the positions and velocities of the point masses composing a particular system. In classical electromagnetism they would be identifiable with the values of the fields at the various spacetime points, and so on. It should also be noted that no hypothesis about determinism or indeterminism is being proposed either: the values of the λ 'of the whole world' at a given time may or may not determine their values at future (or past) times. At the level of generality we have chosen, it is better not to adopt any hypothesis on that particular matter, and the question is therefore left open. Similarly, we leave open the question whether the objective state of a system can be identified with what quantum physics calls the *quantum state* (which, it will be recalled, is the state described by either a wave function or a ket, or more generally by a statistical operator or some other kind of mathematical entity), or whether its full specification requires additional (hidden) parameters.

Given its generality, then, there can be no gainsaying that our working hypothesis about the objective states of systems is an integral part of almost any variety of physical realism. On its own however it is not enough to pick out the salient features of those varieties likely to incorporate the familiar notion of an *object*. Indeed, once we start talking about 'distinct physical systems' we are implicitly raising, albeit vaguely, the idea of separateness and borrowing from everyday experience the concept that certain things are relatively independent of others, notably those distant in time or space. Such independence is of course not an actual fact in all cases. For example, space probes obey commands from Earth quite precisely, even when they are hundreds of millions of miles away. Nevertheless, it generally obtains, if only in an approximate way, and is the very factor that enables the experimentalist to study the behaviour of particular objects, be they physical, biological or whatever, without generally being obliged to take into account what is happening on Sirius.

For the scientist who thinks according to the instrumentalist *principle*, such an independence is just as much a *phenomenon* as any other. It so happens that it holds good. For the *realist* however, the first task is normally to try to integrate such a wide-ranging and important fact into his description of independent reality. Hence it is natural that he should first attempt to explain the possibility the investigator has of disregarding the effects of distant events, by referring to some in-built structural property of *independent reality itself*. In his view, if there is anything like separability in phenomena, the most likely explanation is that this separability inheres in the things themselves.

The true nature of this 'separability' remains to be established precisely. As the counter example of space probes shows, this is not quite a straightforward undertaking.[5] Some precautions have to be taken, and these necessitate that we formulate a 'principle' of separability, with a view to rendering precise what the realist considers at first sight to be 'obvious'.[6]

3 The 'principle of separability'

To be specific let us consider a measuring instrument F, suitable for the measurement of a certain physical quantity A on particles of a certain type (it is not necessary here to propose an hypothesis about the 'real' nature – whether it be a corpuscle, a wave train or both – of the object to be measured; in order to follow the usual conventions, we shall simply call it a 'particle'). Let us assume that F is very far from any other macroscopic object, and that at a great distance from it is a 'source' S of the particles in question – that is, a device which generates and emits them. Let us further assume that at a certain moment F records the passage of a particle U originating from S. Finally let us assume that F simultaneously displays some real number α, said to be 'the result of the measurement carried out by F of the physical quantity A pertaining to U'. We are not excluding the possibility that S may emit, along with U, other particles, perhaps 'in other directions' (the phrase is in inverted commas to indicate that there is no hypothesis of the existence *in itself* of directions in which the particles are emitted). Nor are we ruling out the possibility that one of these particles (which we can call V) might interact with some other measuring device G, distant from both F and S, which measures on V a physical quantity B, with result the number β. Indeed, we are assuming a situation in which both these hypotheses hold good.

We now come to the final but necessary details. Let us suppose that F and G can be given arbitrary orientations, that we can trace the latter by means of two vectors **a** and **b**, and that the nature of the physical quantities A and B actually measured by the corresponding measuring devices depends on the directions of **a** and **b** respectively.

Having specified the nature of the experiment, we may begin to reflect on it. The first thing to note is the obvious fact that anyone unaware of the state of the source S as the moment when U is emitted cannot predict with certainty the result α registered by F upon measuring A. All he can do is indicate the probabilities. Would things be different if (a purely academic hypothesis) he did know the objective state λ of S at that moment? Not necessarily, since we did not assume determinism. In order to remain at a sufficient level of generality we shall therefore consider the *intrinsic probability p* of F producing a particular result when the *objective state* λ of source S is completely specified (as too is **a**, of course). Essentially, the 'principle of separability' consists in declaring that *the value of the number p does not depend, or depends only to a very limited extent, on the direction **b** that the (distant) observer manipulating instrument G may choose to give that instrument* (to avoid ambiguity the case may be considered in which this observer is free to select at whim, immediately before the measurement, this direction **b**) *and that similarly the value of the number p does not depend, or depends only to a very limited extent, on the measurement result then displayed by G*. In a deterministic theory p is 'normally' (that is, if the instrument is reliable) equal to 0 or 1. In a theory assuming intrinsic indeterminism, it falls between these two values.[7,8]

Intuitively, there seem to be strong arguments for the principle. At first sight they even seem compelling. Let us assume the principle is wrong. Such an hypothesis would seem, at least in certain cases, to mean that an experimentalist using F could not disregard what is happening far away (in the region of G) and about which he knew nothing. Consequently, he could have no faith in the information his instrument provides, which would thus be quite meaningless.

We must be aware that this argument is somewhat more subtle than it looks. Its (at least apparent) pertinence is essentially based on the fact that by definition p refers to a *wholly-specified* objective state of the source. If that state were not specified, or were specified only incompletely, the probabilities in question could well be quite significantly different without thereby giving any cause for surprise. Correlations due to common causes at the source would easily account for difference between the probabilities. Thus for example[9] it could well be that the two receptors F and G of two different space probes might react in correlated fashion. More precisely, it is quite easy to imagine that the likelihood of a space probe near Uranus performing a particular manoeuvre at a particular time could be different for an experimentalist unaware of the manoeuvre performed a few seconds earlier by another probe near the Moon from what it would be for an experimentalist aware of what the second probe has just done. For this to be so, the manoeuvres of the two probes need only to be controlled by two wave trains, generated on Earth by manipulating one and the same device. But to regard such an observation as natural and obvious we need to suppose that the experimentalists in question, while knowing the general layout, do not see the details of the manipulation. For if they do, or to be more precise, if they have the exact details of the nature of the individual signals sent to both probes (in particular, of those sent to the one near Uranus) they are then in a position to estimate the likelihood (which, incidentally, will be very close to either 0 or 1 in this 'quasi-determinist' example) *on that basis alone*. To put it another way, for experimentalists so well informed about what is going on at the source, we expect that the extra information about the behaviour of the probe near the Moon will be superfluous and without any influence on the calculation of the probability. Once again it would seem that any other hypothesis is invalidated by the undeniable fact that there are such things as reliable receiving instruments.

If we are to be properly rigorous however we must see that it is necessary to weaken the above argument slightly. A quite proper criticism to make of it is to point out that although the direct influence on the first (Uranus) probe of what is happening on the second (Moon) probe is very slight because of the distance involved, it is nevertheless wrong to argue as if it were strictly equal to zero. Consequently it may be that for example the (very low) probability of an aberrant response from F could be *very slightly* modified by the orientation of the receptor G, or by the response β of G itself, if β is not rigorously determined by the signal sent from Earth (which, within fairly narrow margins, is also quite conceivable). On these grounds it could no doubt be seen as an improper extrapolation from what seems 'straightforward' to insist that *p absolutely does not* depend on **b** and the results provided by G. Nevertheless, the arguments put forward seem to impose a very low degree of dependence. More precisely, they seem to be of such a kind as to convince us that it is always possible – in principle at least – to have the two instruments F and G and the source S at sufficiently great distances from each other for the dependence in question to be measured by a number smaller than some small number fixed in advance. This is what is assumed in a precise formulation of the principle (see [2A]).

In these last few pages essential elements of what we shall now call the 'principle of separability' have been clarified. To state this principle exhaustively, we need to add to the foregoing a subsidiary condition, namely that the probability distribution relating to the objective state λ of the source must be independent of the orientations **a** and **b**. This second condition is quite natural too. In fact if we go back to the example of the two probes, it simply means that the details of the various manipulations carried out on Earth at the command station, which lead to the signals being sent, in no way depend on the orientations **a** and **b** of two far distant receptors each of which may be imagined as localised in the vicinity of one probe and as serving to collect information concerning it.

4 Violation of the 'principle of separability'

As has already been stressed at several points, the 'principle of separability' is the precise formulation of what a realist sees as some kind of virtually self-evident truth. There is no need to insist that in the example of the two probes (as in many others that could be invented quite easily) the principle is fully observed. Indeed it is not violated by *any* of the phenomena of *classical* (as opposed to *quantum*) physics, in either its relativisitic or non-relativistic form.

On the other hand, the 'principle of separability', just as with 'separability in the wider sense', *appears* to be violated in a large number of cases in quantum physics. It definitely is violated if we follow the usual practice of quantum specialists and identify the objective state λ of source S with the entity that they call the 'quantum state' and describe by means of a wave function or ket, or by some more general mathematical algorithm ('density operator', 'linear positive definite functional based on an algebra of observables', and the like). To check that violations then occur it suffices to consider the example of the two particles P_1 and P_2 already used in the first section of this chapter. These have previously interacted and are now each investigated by means of an instrument, these instruments being at a great distance from each other. Here, if we identify λ with the quantum state of the source (F and G being the instruments, with some orientations **a** and **b** which are not important here, and α and β being the results of the measurements of position carried out by F and G on P_1 and P_2 respectively) the application of the quantum mechanical rules of calculation (which in this case have been thoroughly tested and are well established) indeed leads in many instances to values of p that differ widely according to whether the results provided by G are taken into account or not. And this occurs, it should be noted, whatever mathematical algorithm (wave function, ket, density matrix, functional based on an algebra of observables . . .) is used to describe the quantum state of the source.[10]

As stressed in somewhat different terms in some earlier works [17], this is one of the basic differences between classical and quantum physics. At the deepest level of what in theoretical formalism is truly operational (and only the quantum state *is*, while subtler states defined by hidden parameters, even if they 'exist', *are not*), and in so far as quantum physics seems suitable for describing the fine detail of the way matter is built up, matter thus appears to us already at this stage as being basically 'non-separable' (although, as we shall see, phenomena remain 'separable' in every sense of the word that has any connection with human action). It is understandable that many physicists should decide to take their reflections no further than this point. In their view, the considerations just stated, once they have been expressed in more precise mathematical form, will be enough to establish non-separability, and hence the violation of the 'principle of separability'. What they will rightly insist on (as, once again, we shall also do later) is that, in practice, non-separability cannot be used, a conclusion that the formal quantum structure in its actual state is amply sufficient to demonstrate.

But this is where *we* have to go further. For anyone who has stressed, as I did above, the legitimacy of a realist conception of things, the fact that some hidden parameters may be 'non-operational' cannot be enough to rule them out as possibilities right from the start. Now if we refrain from referring to some 'principle of completeness' (which would necessarily be based on a criterion of observability) for denying that these parameters exist, then neither the violation of 'separability in the wider sense' (see the first section) nor that of the 'principle of separability' is established merely by taking into consideration the principles of quantum physics, whatever kind of language – elementary or sophisticated – they are couched in. And then a question arises. Some theorists, not averse to the 'hidden-parameters' concept, consider that there might be a 'sub-quantum medium', the existence of which it would be impossible to demonstrate directly as long as no quantum prediction (concerning some new phenomenon for example) is experimentally falsified, but whose existence would restore strong objectivity. Given what has just been said, are we to believe that in the view of a hypothetical being with access to this sub-quantum medium the 'principle of separability' would be true? *A priori* the hypothesis is far from absurd, for it stands four-square in the tradition of some great discoveries in physics, of which we might take the discovery of molecules as an example. And the principle seems, as we have seen, to mirror quite faithfully man's ability to construct reliable instruments and to study specific objects scientifically. So at first sight trying to 'rescue' it seems a perfectly sensible ambition. As we know however, the physicist, like the explorer, does not always discover what he expects to find, and in a general way this is a cause for rejoicing rather than sorrow. Reality resists us – which proves that it exists. As for the 'principle of separability', there is now a proof that it is false, a proof which in no way depends on any hypothesis about the existence or non-existence of hidden parameters.

The proof in question is based on the fact that the principle of separability entails the validity of certain inequalities (for short called CHSH, after Clauser, Horne, Shimony and Holt) which generalise inequalities due to Bell and which bear on certain measurable (and indeed measured) quantities. The rules of calculation in quantum mechanics imply that, in some cases, these inequalities should on the contrary be violated. There is thus a possibility of an experimental test.

This constitutes a considerable advance. As things were before, it would have been possible to decide for or against the 'principle of separability' (and in fact such decisions were taken for or against similar views) only on the basis of *a priori* philosophical options qualitatively comparable (no irreverence intended) to those which in antiquity led some to declare that the Sun is bigger than the Earth, while others thought it smaller. In the particular case we are discussing, those taking an operationalist view of things would have declared the principle to be violated (or to be meaningless), while those with a realist turn of mind would have preferred the opposite position. That indeed is roughly what happened, even though the principle in question was not formulated as precisely as it is today. Today, just as science enables us to decide clearly about the relative masses of the Sun and the Earth, so too does it provide, in principle and to a large extent in fact, the means of deciding whether the 'principle of separability' is true or false. The decision turned out to be in favour of its falsity. Given the experimental data, the CHSH inequalities are found to be violated.[11] What is more, quantitatively they are violated in exactly the way quantum mechanics led us to expect.[12]

As is known, this conclusion rests on both theoretical deduction and experimental measurements. The former enables us to show that if the principle of separability were true, the Bell–CHSH inequalities would necessarily be experimentally verified. In the most general case, the deduction in question entails a very easy but rather long calculation (which will not be reproduced here; the details can easily be found, in [2A] for example). Fortunately, there is a particular case (that of 'strict correlations', as discussed above) in which the deduction can be made without reference to any mathematical technicalities, even of an elementary kind. The principle involved has been described in ISR [2] and in [21].

As for the experimental measurements, they consist of noting for each pair of particles (U,V) the numbers α and β defined above as the results recorded by instruments F and G respectively. (In general, these latter have possible values of 1 and -1, in appropriate units.) This elementary operation is repeated on a very large number of pairs, the instruments being oriented towards directions $\mathbf{a}$ and $\mathbf{b}$, and the number $N_{a,b}$ of cases in which the two results α and β are of the same sign is noted. The above sequence of operations is performed three more times, the instruments being oriented towards $(\mathbf{a},\mathbf{b}')$ the second time, $(\mathbf{a}',\mathbf{b})$ the third time and $(\mathbf{a}',\mathbf{b}')$ the fourth.[13] The Bell–CHSH inequalities bear on the numbers $N_{a,b}$, $N_{a,b'}$, $N_{a',b}$, and $N_{a',b'}$ thus obtained. In practice, particles U and V are usually photons, the physical quantities measured are rectilinear polarisations along the directions $\mathbf{a},\mathbf{b},\mathbf{a}'$, and $\mathbf{b}'$, instruments F and G are rotatable combined polarisers, analysers and detectors. (For further information interested readers are referred, for example, to [20].) Here there is no need to go into further detail about the experimental apparatus.

Suffice it to say that such experiments have been repeated independently several times in large university departments in both Europe and America, that they involve some of the most advanced techniques available to physicists (and statisticians), and that their results have led to the almost-unanimous conclusion among the international community of physicists that the inequalities in question have effectively been violated, as the quantum rules predicted they would. This means the idea that this 'violation' could be attributed merely to imperfections in experimental technique can no longer be seriously maintained.

To be completely objective however, we should mention that there are still one or two groups of theorists (very much in the minority, it is true) who believe they can maintain the opposite of what has just been said. In their view, neither the theoretical nor the experimental results reported above are yet quite powerful enough to force us to give up certain 'common-sense truths' still considered to be the backbone of classical physics. Beneath all the differences that divide them, these theorists (and only those whose reasoning is correct are considered here) share a common fall-back position. The great difficulty in this matter, they stress, is to obtain perfectly 'pure' verifications of the (in principle observable) predictions entailed by a theory. They grant that, in the case in point, not giving up the 'common-sense truths' in question (of which the details are not significant here but separability would be one) implies that some of the predictions arising from quantum mechanics and which are in principle verifiable are in fact false. They do however stress the great difference between 'in principle' and 'in practice'.[14] No instrument is perfect. No meter records *absolutely all* the events it is supposed to record, and the records of a detector, however great the care with which it has been constructed, may always be influenced by its past history, and so on. They see it as conceivable that, for certain reasons we cannot at present grasp, all these flaws[15] in fact *conspire* to make the results actually registered by the instrument statistically compatible with the predictions entailed by the 'strictly speaking false' theory that, in their view. quantum mechanics is. (Here 'strictly speaking false' means that if, which is not the case, all the predictions the theory entails could be tested with instruments having none of the flaws we have listed, discrepancies between theoretical predictions and experimental results would indeed emerge.)

The idea of a conspiration of such a kind seems to me – and to the majority of theorists – a great deal more unpalatable than the idea of abandoning the principle of separability. It is true of course that in the history of physics some such 'conspiracies' (one thinks of the Lorentz contractions 'conspiring' to make the motion of bodies relative to the ether unobservable) were in fact subsequently found to have seemed unbelievable only in the absence of a theory truly adequate to the data. In such cases however, the new theory (the theory that did remove the semblance of 'conspiracy') was never aimed at, nor resulted in, salvaging some common-sense ideas or prejudices. But that is precisely what the theorists discussed here implicitly expect of some theory yet to be produced when they appeal to a justification of such a kind. In the language of contemporary epistemology, we could say that while making an existing theory immune to criticism is legitimate if the immunisation leads to verifiable predictions (see the example of Uranus and Neptune to be discussed in the next chapter), on the contrary the purely qualitative 'immunisation' of old ideas (by systematically appealing to our ignorance or the imperfections of our instruments to hide the clashes between the consequences of such ideas and the predictions of a theory which is otherwise successful) is definitely objectionable.

Conclusion

If we agree to define non-separability as the violation of the 'principle of separability', then given the foregoing developments we can and must maintain that non-separability is proven. Clearly such a conclusion is of considerable importance. However we should be careful not to read more into it than is actually there. Here, once again, we must keep our critical faculties fully alert.

Contrary to first appearances, one implication we cannot read into the above conclusion is that it is possible to send usable long-distance signals which would be propagated instantly, or at any rate faster than light. If, as everything seems to show, the verifiable predictions – the rules of calculation – of quantum physics are correct it can be shown (see for example [2A] p. 248) that non-separability cannot provide any possibility of long-distance human action operating at a speed faster than light (which refutes the hasty and sometimes rather noisy speculations about useful applications we occasionally hear). It is just as important, however, to see clearly that trying to use that observation as a basis for reducing non-separability to a characteristic of known classical effects would be just as illusory. We should not, for example, suppose that non-separability would be appropriately illustrated by the speed, greater than that of light, at which a spot created on a (notional) circular screen a million miles in diameter by a central revolving light would move. An effect of that kind in no way violates the realistic-relativistic axiom that all the *physical consequences* of an event are contained within its future light cone. The case is very different indeed when the 'principle of separability' is violated, for this does involve a violation of the axiom every time – and these are the most interesting occasions – the two 'measurement-events' considered above are separated by a space-like interval and the instruments are positioned just before they are brought into operation. Even if the violation does not bring about any possibility of action, it is nevertheless disconcertingly novel to the classical realist, quite unlike the moving spot mentioned above.

Another conclusion that it would be premature to draw from what we have seen in this chapter – because it does not follow from non-separability alone – is that here we are dealing with a strict refutation of microrealism in all its varieties, including those furthest removed from the classical conception of things. Strictly speaking, it is possible to imagine that there may be theories that are *literally* microrealist, but in which distant forces of very strange behaviour might entail non-separability. Indeed, we should note that such theories have in fact been constructed. Because they are deterministic as well, they will be discussed in the next chapter, which deals with that question.

In connection with them, we can however already point out two things. The first is that, for reasons which will be explained in that chapter, such theories have not received the endorsement of the majority of physicists. The second is that such a kind of 'non-separable' microrealism is by that very fact robbed of all the major features that *a priori* make microrealism an attractive way of conceiving reality. For example, in the case of correlations at a distance considered a little earlier, the chief virtue of the kind of microrealism we were then speculating about was that it was expected to account for those correlations by means of an explanation in terms of 'common causes at the source', an explanation fully analogous with the current explanation of macroscopic correlations of the same type (see the example with the probes) and which thus seemed quite natural. But the violations of the CHSH inequalities ruled out such an explanation, and the non-separable microrealism puts in its place a quite different one which does not have the same grounds for its apparent support.

6
Physical realism in trouble

Even when microrealism has been discarded, some less particular versions of physical realism remain *a priori* tenable. As the title indicates, this chapter offers an account of the refutations which have been made, or which I think need to be made, of some ideas which, were they well-founded, could revive the realist conception in some more general form. Since these ideas might, if we did not examine them as I propose to do here, seem simple, attractive and true, it is important at this point to show that they are *false*. The fact that *a posteriori* they are untenable is one of the most powerful arguments for trying to find different ones, a task that will be undertaken in Part III of this book.

On the other hand, refutation is necessarily an arid process. It is often exciting to encounter new ideas, but to see other ideas come up against difficulties that ultimately are insurmountable is rarely so. Fortunately, although such a detailed critical examination (presented in a very summary form here) is necessary, the reader may, if he so wishes, leave a close study of it until later. A first reading of this chapter might fairly safely be limited to skimming through it (leaving aside altogether sections 5 and 6 if time is short).

However, before we begin looking at this kind of refutation, we should be aware of another one that has already been outlined. It is concerned less with physical realism as such than with a question of how to interpret a particular aspect of the information given above.

1 The problem of influences at a distance

In the last chapter we saw that any realist conception of the world which accepts the validity of the principle of separability is erroneous. This raises questions of the following kind. Must we then conclude that there do exist forces which act instantaneously at a distance? That there are *causal relationships* between events that are at once so distant in space and so close in time that even light would not have time to carry a message from one to the other? Must we in fact talk of influences at a distance propagating faster than light? Unfortunately, such questions are more complex than they seem, since the way they are formulated involves terms such as ‘force’, ‘causality’ and ‘influence’ whose ordinary definitions are not precise enough to allow us to master the formidable problems that arise in this context. We find ourselves, in effect, in the same difficulty we encountered earlier. To express our thoughts, we need words, and in fields as unfamiliar as this one the meanings of everyday words need to be defined precisely; but this precision is hard to attain without recourse to definitions of an operational nature. The idea of causality, for instance is nowadays generally defined by physicists on the basis of at least conceivable (human) action. If we adopt that kind of definition, we can answer the questions raised above only in the negative. Indeed – to repeat – it is possible to demonstrate the following. The fact that the verifiable predictions of relativistic quantum mechanics are correct (and up to now they have all proved to be so) is sufficient to prevent us from deriving from the violation of the 'principle of separability' any possibility of human action operating at a distance faster than light. On the other hand, if we consider (which on the part of the realist is far from unreasonable) that whatever the (perhaps still to be defined) meaning of the term 'influence at a distance', it must in any case be such that a violation of the 'principle of separability' corresponds to the existence of such an influence, then we must grant (a) that (unless we imagine some kind of purely *ad hoc* 'conspiracy' of the kind previously discussed) such a violation is in fact observed, and (b) that it *does* imply the actual existence of influences that operate faster than light. Therefore the questions we are talking about must necessarily be answered in the positive. . . . Unless, of course, we engage in efforts to distinguish between the ideas of cause and influence, which is clearly a delicate undertaking.

In order to avoid the obscurities and pitfalls that always attend debates about what words mean, the problem of whether influences operating faster than light do or do not exist will not be discussed here in fully general terms (see [2A] for further details). All we need do is remember that contrary to what was thought in the era of classical physics, the 'principle of separability' is indeed violated.

2 The problem of determinism

Of course we cannot expect this violation not to affect the question of determinism. Indeed, we shall see that it plays a role in every model we care to specify, and its inevitability should make it plain from the start that if some determinist theories *are* admissible, they can never be analogous to the *classical* ones which provide correct large-scale descriptions of the motions of planets, of billiard balls and so on. But having duly noted the warning, neither must we take as our starting point the essentially negative notion of non-separability. It is more appropriate to study the problem of determinism on the basis of ideas more closely related to the main traditions of rational thought.

One such idea, which though perhaps difficult to justify is in fact widespread, is that there should be quite a direct link between realism and determinism, that each should imply the other. The idea is hard to justify because although it is hard to imagine a deterministic theory that would not be 'realistic' (by which I mean: not purporting to be at least a partial description of 'what is'), the opposite is in fact not hard. I can imagine a world in which on the one hand localised things would not only be *objects for a subject*, but also *objects in themselves*, and in which, on the other hand, a pencil (an object in the latter sense) balanced exactly on its point would fall in a truly random direction, that is, a direction not predetermined by any known or unknown parameter. Whether the real world is in fact constituted in such a way is of course quite a different question. What I wish to stress here is that there is no logical flaw in making this separation between realism and determinism.

Nevertheless, it is a fact that the two ideas are almost always associated with one another. They were, for example, by Spinoza and by Einstein. The latter's famous declaration that he could not believe that God plays dice with the cosmos was – it should not be forgotten – made at a time when one of his chief aims was either to *change* – or as he himself said – to *complete* quantum mechanics in order to make it a theory interpretable as a true description of reality rather than as a mere tool, however effective, of prediction. The reasons for the association in question are of course at least in part psychological and aesthetic in nature, but that does not mean we should ignore them. For a consistent realist, there is nothing 'greater' – in every sense of the word – than the Universe, and it is therefore understandable that he should find such 'greatness' incompatible with the idea that a major part is played by chance (or 'noise', as determinists such as René Thom [22] prefer to say). To this it should be added that in most of the cases in which we invoke the part played by chance it is simply a question of the coming together of two easily traceable causal series having no obvious common origin. Clearly this kind of 'chance' is compatible with the deterministic hypothesis.

Be that as it may, for these and other quite legitimate reasons the determinist thesis is, even today, put forward by a number of leading thinkers as a kind of prerequisite for rational thought. They see it as perfectly defensible provided only that we are willing to pay the price involved, namely, given the data of present-day physics, to apply the thesis only to a level of reality somehow deeper than what is directly – or even operationally – observable. But in the next few sections we shall see that even when thought of this way, the determinist thesis remains very hard to accept.

What complicates the examination of the thesis and the analysis of its implications is the fact that it is applied to contemporary microphysics in several different ways, each of which needs to be studied separately. Fortunately, these ways fall essentially into just two groups. Those in the first are based on the idea that quantum indeterminism is ultimately only an *apparent* indeterminism, which arises solely from our ignorance of *fine details*. We should therefore attempt to complete quantum description by assuming the existence of additional ('hidden') parameters, usually differing from one physical system to another even when both have the same wave function. Together with the wave function at a given point in time, these parameters (which were subsumed under the symbol λ in the last chapter) determine the future development of the physical system they pertain to, and it is only because we are ignorant of their values that this development seems to us not fully determined.

The other group of attempts to restore determinism includes those based on a feature of quantum physics which is indeed important and suggestive. I have in mind the fact that as described by Schrödinger's equation (or the algorithms replacing it in the relativistic formulations of quantum mechanics) the development *of the wave function* is determinist. Given that the wave function (or, here again, the more elaborate algorithms replacing it in the relativistic formulations) is the essential element of the fundamental mathematical formalism underpinning the whole theory, that fact alone might at first sight seem sufficient to 'prove' the deterministic nature of this theory. However, as we shall see later, things are not as simple as that.

3 Hidden parameters and determinism

It is well known that statistical thermodynamics, a science founded in the nineteenth century, in large measure by Boltzmann, makes great use of the idea of probability. The probabilities its most common forms make use of, however, are subjective; that is to say, they refer to our lack of knowledge. Basically, Boltzmann and other proponents of the theory assumed that the molecules whose existence they postulated obeyed the laws of classical mechanics, which is a determinist theory. At the time Boltzmann was working, these molecules could not be detected experimentally. Besides the justification for introducing the concept of such molecules (or indeed that of atoms, which is very closely akin to it) was in fact purely theoretical and even speculative: such a notion made it possible to construct a theory which related thermodynamics as it had existed so far to other areas of physics. But later developments in physics showed that the idea was indeed most fruitful, since it suggested the existence of a great number of hitherto unknown phenomena, which experimental evidence established that they do in fact occur as has been predicted theoretically. Now, as can be seen, the hypothesis that there are extra parameters not described by wave functions, and re-instating the determinism called into question by the usual quantum interpretations, is obviously quite similar to Boltzmann's theory of the existence of molecules. Indeed, it should enable us to reduce quantum probabilities to the status of subjective probabilities; and just as with Boltzmann's theory, it should lead to knowledge of a deeper level of nature, which is often described by its proponents as the subquantum level. It would seem quite legitimate to expect an idea of this kind to be fruitful.

From a qualitative point of view therefore, such a programme seems eminently reasonable. But since all science is *quantitative*, what we must now consider is whether and to what extent the general idea just described can take shape within a precise theoretical structure.

In so doing, we must first beware of unjustified objections, of which there are a number. Their common feature is to introduce arguments of a practical or operational nature into a question which by definition is not subject to them. The archetypal example of this kind of confusion is the objection based on Heisenberg's relationships, often referred to as the 'uncertainty principle'. This objection maintains that the very existence of these relationships rules out determinism. There is obviously no basis for this objection since, as we have seen, the 'uncertainty principle' could in fact be just a matter of *uncertainty*. In a vessel full of a gas assumed to obey the laws of classical mechanics, there is no question of our actually being able to *know* where in the vessel a molecule whose position is well defined at the moment will be in a few minutes time even though determinism is postulated. In the same way, from the sole fact that the position and velocity of an object are not *knowable* simultaneously (and with a precision exceeding a given limit), we cannot *conclude* that *in itself* the object in question does not, at a given moment, have both a well-defined position and a well-defined velocity, which acting in concert with other known or unknown parameters, might well determine its subsequent behaviour.

Another equally unjustified objection is the claim that the very existence of hidden parameters would make what are known as quantum superposition phenomena, in particular the interference phenomena predicted by quantum mechanics and actually observed, quite impossible. It is true of course that it is far from trivial to interpret the results of interference experiments on particle beams in terms of hidden parameters and deterministic behaviour. This is because in Young's slit experiments, for example, an interpretation of that kind necessarily requires that the behaviour of the particle passing through *one* of the slits be affected by the state (that is, whether it is open or closed) of the *other*. To be sure, anyone can legitimately maintain that this introduces quite an awkward complication into our theoretical view of things, and that therefore the conventional indeterministic conception is preferable. However it is clear that such a line of argument alone is unlikely to make a thorough-going determinist change his mind; he can always argue that there may exist certain yet unknown fields which induce just such influences.

To what extent can such a reply be considered satisfactory? Philosophers would probably (and in one sense quite rightly) recommend us to think here in terms of 'immunisation' (or of the related concept of the 'protective belt' surrounding a theory, described for example by Lakatos [23]). As we already know, contemporary philosophy of science has brought to light the role and importance of this procedure. It consists in introducing a supplementary hypothesis by means of which the scientific community attempts – sometimes but not always successfully – to preserve a theory hitherto deemed satisfactory from the onslaught of some specific objection, for example one based on a new discovery. The history of science abounds in immunisations of all kinds, and their detailed study makes it possible to isolate two conditions apparently necessary for the process to work. The first is that neither the simplicity nor the overall synthetic power of the initial theory should be adversely affected; the second is that the supplementary hypothesis should be productive, in other words that it should predict new facts which are confirmed empirically.

The attempt to immunise the geocentric Ptolemaic theory by introducing epicycles provides an example of an undertaking that ultimately proved unsuccessful with regard to both these conditions. Introducing the new concept obviously complicated the theory, and history tells us that it led to no empirically-confirmed predictions. In such a case, it is appropriate to speak of *a purely* ad hoc *hypothesis*, with all the pejorative overtones of that expression. To set against it we have an example of a successful immunisation in the discovery by Le Verrier and Adams of the planet Neptune by studying anomalies in the trajectory of Uranus. In the first case, observations of the planetary motions contradicted the basic hypothesis of geocentricity; (hence the attempt to immunise the hypothesis using the idea of epicycles). In the second, similarly, anomalies in the trajectory seemed at first to contradict Newtonian theory. But unlike the immunisation in the first case, in the second case the hypothesis of the existence of a new planet genuinely provided a way round the difficulty, for it met both the conditions set out above. Not only did it in no way diminish the simplicity of Newton's theory; it also demonstrated both the predictive power of the latter and its own productiveness, since calculations founded on the combination of the new hypothesis and the basic rules of the original theory led to the discovery of Neptune.

There would seem to be no reason why the conditions relating to the immunisation of scientific *theories* should not also be applied to the immunisation of the *interpretation of formalisms*. In particular, it may seem appropriate to apply them to the attempted immunisation of determinism, which the hypothesis of hidden parameters (or 'variables') in quantum physics amounts to.

If we follow this path, we should of course note in the first place that the merely *qualitative* hypothesis that hidden parameters exist meets neither condition. However, such an observation is scarcely compelling by itself. We must bear in mind that deterministic models that are genuine hidden-variable *theories* have in fact been constructed. One of the oldest (and best) was devised by Louis de Broglie [24], shortly after his discovery of the wave properties of matter. This mathematically well-defined model, subsequently re-invented and refined by David Bohm [25], is sufficiently general for its experimental predictions to coincide both qualitatively *and* quantitatively with the corresponding predictions of standard quantum mechanics, at least within the substantial domain of validity they have in common.

One can, after a fashion, defend the view that such models are satisfactory immunisations of the determinist thesis. The arguments in favour of this point of view arise from the generality and the quantitative nature of the models in question, two features which guarantee that if in the domain covered by elementary quantum mechanics any phenomenon whatsoever is studied and the models provide usable information about it, that information will be identical to the data derived from the standard indeterministic model. In other words, we can say that the first of the conditions set out above is met, at least in part. The second however is clearly not. The models in question have now been known for more than half a century and within that time they have led to no new predictions in the empirical domain; otherwise said, they have been quite unproductive. Compared with the discovery of Neptune by Le Verrier and Adams, this fact may well induce us to enter serious reservations about them.

'Reservation' however must not be taken as amounting to 'rejection'. In effect, the condition of productiveness, when we think about it, is plainly an instrumental condition. Though essential when used in considering the legitimacy of immunising a *theory*, taken in the usual sense of that term, it may with good cause be questioned when used to test an *interpretation* of a theory. Hence, at the present stage in our proceedings, it is not yet appropriate to declare the objection summarised above to be absolutely decisive.

That being so, we need to extend our analysis. And at this juncture, as we might well suppose, we must bring back into consideration the matter of non-separability. As we have seen, non-separability requires that the correlations we have been led to expect by quantum mechanics and have observed in the experimental results from pairs of particles (Chapter 5) should be accounted for *in some other way* than by correlations established at the source between the parameters of both particles (since the latter attribution would necessarily imply inequalities which are violated). Within any deterministic schema, the results recorded at a particular point in time by one of the instruments must therefore be at least partly determined by one or more of the events which take place a great distance from where the measurement is made and which are directly related neither to the instrument carrying out the measuring operation *nor* the particle undergoing it. It is hard to deny that this fact already very much reduces the simplicity and elegance of the deterministic hypothesis, and so greatly diminishes its attractiveness.

But there is more. Within the framework of certain kinds of experimental set-up designed to create conditions equivalent to those obtaining when instruments are very mobile (and if we once again reject as purely *ad hoc* the hypothesis of highly improbable 'conspiracies' between the hidden parameters of these instruments), the events referred to above are so near in time and so distant in space from the 'measurement-event' in question that any 'influence' they might exert on the latter must necessarily travel faster than light. Is this a violation of 'Einsteinian causality'? Not if we mean by that expression the impossibility of transmitting a *usable signal* faster than light, since as we have already seen, these experiments enable us to do no such thing (which can be rigorously demonstrated and is mainly due to the fact that we cannot ourselves choose the results of our own measurements). So we can see that *a priori* it is possible to envisage deterministic theoretical models that would both take into account such data and be compatible with Einsteinian causality understood in this 'instrumentalist' sense. Some theorists are currently trying to proceed in just this direction (see for example Chapter 10), and their efforts deserve serious attention. But others may want to go still further in the direction of realism, and this will lead them into quite basic difficulties. In particular, it is not enough to construct an abstract schema that allows for influences which travel faster than the speed of light but which do not permit the transmission of signals. The influences in question must also be such as to provide a detailed account of the specific correlations which, via the line of reasoning set out above, led to their introduction in the first place. Now in just the kind of experimental set-up we are interested in here (we may think for example of the one used by Aspect *et al.* [20]) the correlations occur between measurement results recorded by two instruments between which, *in relativistic physics*, there is complete symmetry. This is because in relativistic physics, as distinct from non-relativistic physics, it is not possible in such cases to specify in an absolute way which measurement-event occurs first. That depends on the frame of reference, which is to say on the 'overall observer'. The influence relationships we are led to postulate between the two measurement-events cannot therefore be indicated by a directed arrow, since in a sense each is both 'cause' and 'effect'. One may legitimately ask whether such a non-directed influence relationship is *truly* an influence relationship; and if it is not, what is its status? The problem looks thorny, and indeed, though some claims have been made, there does not at present seem to be any deterministic *and* basically relativistic theory genuinely able to provide a detailed explanation of the correlations we have been discussing.

We can see that seeking a conception of the world that is at once realistic, deterministic and relativistic is a perilous undertaking. We can sum up what has gone before by noting that of the two conditions that philosophers argue a successful immunisation process must fulfil, one (productiveness) is clearly not met by the determinist hypothesis examined here, whilst the other (simplicity and synthetic power) is not truly met at present and may well never be unless we understand determinism in a totally new way. This state of affairs goes a long way to justifying the scepticism most contemporary physicists feel about the deterministic interpretation by means of hidden parameters that we have been looking at here.

4 Determinism and the wave function

We might think that, rather than constructing highly complex models which – like those we have just examined – from the outset depart from the spirit of quantum mechanics (since they reproduce only artificially the superposition effects that are its essential feature), we should rather let ourselves be guided by the deterministic character of the temporal evolution of the wave function. We all know that the wave function or ket obeys Schrödinger's time-dependent equation, or its relativistic equivalents. The value of the wave function at all places and at all points in time is thereby determined once it is known at all places and at a single point in time. After a fashion, we can therefore maintain that quantum mechanics is *already* a deterministic theory and that consequently the problems we are examining in this chapter simply do not arise . . . .

This argument is certainly attractive. Of course it implies – since determinism in fact presupposes physical realism, as we have seen – that the wave function is a physical reality, and more specifically that it is what *constitutes* reality; or (what is more or less the same thing) that it gives us a complete and faithful picture of reality. Of course, seen in this light reality can only be 'remote' (as opposed to 'near', in the sense I ascribed to this word earlier, see Chapter 1 p.24). In other terms, reality cannot be reduced either to a 'wave' in the usual sense of the word, nor indeed to anything that can be described using everyday concepts. For example, the wave function of a system made up of several particles cannot in general be described by giving the values of the wave functions of all the particles at every point, since it is generally impossible to assign a specific wave function to each particle in the system. To a thorough-going realist (as our determinist necessarily is) this is not however a decisive objection. At a pinch, it is always possible to imagine reality as occupying not the four-dimensional space-time we can perceive but rather a space with considerably more dimensions, of which we can perceive only certain projections.

Be that as it may, we still need to take our analysis further. In doing so we must expect to encounter certain difficulties when we come to the description of events linked to the process of measurement. In its usual form, the quantum formalism involves not *one* law of evolution, as might be expected, but *two*. One of these – the problematic one – is appropriate for describing phenomena in which a physical quantity is measured by means of an instrument.[1] More precisely, the formalism applies Schrödinger's (deterministic) equation to the evolution of the wave function of a system *only when the latter has not been subjected to any measurements*. To describe it after measurements have been made, and specific results obtained, a *new* wave function based on calculations incorporating those results is introduced. This operation, known as the reduction (or 'collapse') of the wave function, generally produces a new wave which is qualitatively very different from the former one, and the transition from the first to the second at the moment the measurement is made is not described in Schrödinger's equation. In fact, although this equation describes a deterministic evolution, the transition in question is indeterminate. Like the results of the measurement itself, the second wave function is predictable only in probabilistic terms.

Do the facts we have just been considering constitute in themselves a decisive objection to the thesis? In all honesty, we cannot at this stage claim that they do. The quantum formalism offers what might seem to be a suggestion for an answer. The point is that the measuring instrument itself could well be treated as a quantum system (a way of seeing things that at first sight even seem inescapable; is not every object made up of atoms?). If we adopt this point of view, the interaction between the microsystem and the instrument now becomes merely a particular example of the interaction between two quantum systems. Taken together, they form a composite system S, with a wave function that develops in accordance with Schrödinger's equation, that is to say, in a deterministic way.

Is this the definitive solution? We cannot risk accepting it as such until we have solved an enigma, which can be formulated thus: what becomes of the *probability* of this or that result of measurement, which the usual quantum formalism enabled us to calculate? If the formalism generates true predictions, how is it that the probabilities in question no longer arise? And on the other hand, how *could* they, since determinism is re-established without there being any need to admit the existence of parameters, on our ignorance of which these probabilities could (subjectively) be based?

The answer lies in part in the fact that, when calculated on the basis of the Schrödinger equation the wave function of the composite system S after the measurement interaction has in general a rather complex structure. As we have seen, it takes the form of a sum of several terms, as many in fact as there are *a priori* possible measurement results, given the structure of the apparatus. Apart from a numerical coefficient, each of the terms is the product of an 'instrument wave function' and one possible wave function of the microsystem being measured. Each of the 'instrument wave functions' describes a possible state of the instrument. For example, if the instrument has an indicator giving a numerical reading of the result of measurement, each of the 'instrument wave functions' corresponds to *one* of the numbers which *a priori* could appear on the indicator, and gives an adequate description of the state of the instrument in all the cases in which that particular number is displayed. As for the associated 'system wave function' (within the product being considered), it is the one that would describe the microsystem in question (the microsystem being measured) if the physical quantity which the instrument is designed to measure on this microsystem actually had a numerical value equal to (or uniquely related to) the number in question. Finally, the numerical coefficients which we mentioned in association with the considered product have a significance which, as regards the relation between the mathematical formula described here and empirical observation, is very important. The magnitude of each of them enables us to calculate the probability with which the usual predictive formalism (purely operational, as described above) forecasts that we will observe on the system the corresponding value of the physical quantity being studied.[2,3]

So we see that, in the deterministic theory, the mathematical structure of the wave function of *S* after measurement does in a real sense contain the probabilities of the indeterministic theory. But it does so, one might say, only in 'virtual' fashion. If, in accordance with the hypothesis we are analysing, the wave function of the composite system *S* provides the sole description of the physical reality of that system, what we must say is that *in reality* *S* is in a state which is not the one that corresponds to some definite value of the number displayed by the instrument; rather, it can be described in no other way than as a *quantum superposition* of all these states. Nor can we try to interpret the phrase 'quantum superposition' in terms of familiar classical concepts. It simply cannot be understood in this way, since it refers solely to the mathematical structure, as described above, of the wave function, a structure which is a superposition – or linear combination – of terms.[4]

We now see the price to pay for re-establishing determinism by means of this approach: we must accept that even macroscopic systems such as our composite system *S* can be in physical states which our mind cannot conceive of, and which are 'superpositions' of macroscopically different states. This conceptual price is already so high that most physicists prefer to abandon the deterministic thesis. They stress that when we look at an indicator display which is part of a composite system *S* we always see it in a particular state corresponding to a quite specific number (or at worst to the transition between two consecutive numbers on the moving dial) and never (which is unimaginable) in a 'state of quantum superposition', in which all the numbers on the dial would appear at once. This, they maintain, means that the determinist thesis at hand leads us to expect a situation that is absurd, so that the thesis is therefore clearly not tenable.

There are dissenting voices, however. The objection we have been considering is not seen as conclusive by some, who reply that after all, even if the superpositions of macroscopic states do not correspond to our direct experience, does that prove they *really* do not exist? But such people cannot restrict themselves to just that answer, even if it is supplemented by the remark that observing the dial is never a direct act of awareness because it is performed, for example, by means of the human eye, which is also in some way an instrument . . . and so on. Whatever account of this chain of successive 'instruments' we produce, it is ultimately always the case that there *are* acts of awareness on the part of observers, and these always consist of quite specific impressions, such as seeing a particular number. If such impressions are themselves 'physical facts' and if quantum mechanics – in its deterministic interpretation – is presumed to apply universally (at least as far as its general principles are concerned), it must take this characteristic into account. This leads dissidents to adopt the very bold hypothesis of a branching Universe, also called the 'relativity of states' theory [26]. In this scenario, the observer himself (and his 'awareness') are included in the macroscopic system we called 'the instrument'. Consequently, the 'act of measurement' becomes non-specific. There is only one law of evolution in nature, the deterministic one Schrödinger's equation calls for. But a further consequence is that even the observer's state of awareness after measurement, which is a component of system S, is now divided into as many parts as there are numbers on the dial (to keep, for simplicity's sake, to the example of the experimental apparatus already used).

All this leads to an inescapable conclusion. In a certain sense, all the possible states of awareness (seeing number one, seeing number two, and so on) exist simultaneously. To some degree, the observer is 'spread out' over several mutually incompatible states.

Clearly, however, the question arises of whether the last two sentences (particularly the second) are meaningful or simply rhetoric. There are those who, in order to make them meaningful, would not hesitate to claim that when a measurement or other similar events take place, the Universe is actually enlarged. For them, the fact that a measurement has been carried out means that there is no longer a single observer, as was the case immediately before the operation, but as many (originating in him) as initially there are readings of numbers on the dial of the instrument. (After the operation there would of course also be as many systems and as many instruments.) There are other theorists who believe that it is not necessary to go to such lengths to keep the sentences meaningful. . . .

What are we to make of such views? Radical judgement would be easy if such theories predicted numerical experimental results which would be different from those predicted by quantum mechanics and which could be tested by observation. That however is not the case, as can be shown. On the other hand, the fact that the 'interpretation' closely follows and indeed is wholly shaped by the structure of conventional quantum theory can be seen as a positive feature, even if the resulting statements seem strange. After all, we might think of a parallel with Einsteinian relativity, which from a certain point of view can also be considered as an 'interpretation' modelled on mathematical formalisms (such as Maxwell's equations, Lorentz's formulae and the work of Poincaré) in large part prior to it. The strangeness of Einstein's theory and its shock to common sense did not prevent its wholesale acceptance by physicists. On the other hand, it is worth pointing out that relativity was adopted precisely because it was productive. In other words, it is not *just* an interpretation, but also a 'genuine' theory in that it makes it possible to predict and systematise a great number of facts discovered after it was devised. For the moment, the theory of the 'relativity of states' has not reached that stage, and few theoretical physicists would care to predict that one day it will.

The minority who *do* think it will are mostly theoretical cosmologists, which is quite understandable. Whereas the idea of basing quantum physics on the concept of taking measurements (as developed by the Copenhagen School) is conceivably defensible, even for a realist, as long as it is applied to the microscopic, it begins to look for that same realist rather like false reasoning when there is an attempt to apply the laws of quantum physics to the description of the Universe, since by definition it is impossible to imagine any measuring device not included in the Universe. I hasten to add, however, that so far such general ideas have not crystallised into a truly systematic theory of the Universe taking them as its starting point. Existing theories – often popularised – of the Universe are in fact a mosaic of theories, each stemming from a stock of ideas which differ greatly from one theory to another. So general relativity, for example, which up to now has a *classical* basis only, rubs shoulders with the quantum theory of fields as well as with that classical-quantum hybrid, statistical mechanics. It would therefore be premature, to say the least, to take the theory of the relativity of states quite seriously.

Clearly it is very difficult to sustain the determinist thesis. In the sections which follow, we shall move on to a short examination of two suggestions often advanced, without any deterministic presuppositions, in an attempt to salvage physical realism while at the same time preserving the hypothesis that the quantum rules for predicting phenomena are universally valid. The first suggestion is that of a realism based to an even greater extent than earlier kinds of Pythagorean realism on a reification of mathematical concepts of a very high level of abstraction. The second, more conventional, is to aim at a *macrorealism* which, by referring to certain properties of very complex quantum systems, would reconcile the just mentioned hypothesis with the notion of macroscopic objects fundamentally governed by classical mechanics (with all the latter's recognised strong objectivity). We shall see, however, that neither seems likely to achieve its aim.

5 Strong objectivity versus quantum physics

As we have already noted, what is truly 'solid' in quantum physics is the ensemble of its *rules of prediction*. The weak objectivity (intersubjective validity) of the rules of quantum mechanics is hard to dispute, and is enough to make that discipline the prodigiously effective tool for synthesising our empirical knowledge that it is today. In a sense this is fortunate, for in fact it is very difficult to express some quantum principles in terms of strong objectivity. This is particularly true of the principle that furnishes probabilities, the detailed expression of which posits that when a physical quantity A is measured, *the probability that one, a_j, of the values the quantity may take should actually be the observed one is the square of the modulus*[5] *of the numerical coefficient c_j of the function f_j corresponding to a_j in what mathematicians call an 'expansion*[6] *of the initial wave function' f on a set of functions $\{\ldots f_k \ldots\}$ associated with A in a well-specified way.* There is no denying the objectivity of this statement, since it holds good whoever is assumed to be taking the measurement. But the objectivity is no more than *weak*. In fact, the statement refers specifically to an operation carried out by human beings, an operation which also involves a macroscopic physical system whose essential feature is that it is a *measuring device* – something which by definition is a tool intended to increase human knowledge.[7]

No doubt it will be conjectured by some that this presents no difficulties, and that we can restore strong objectivity by expressing ourselves differently. By analogy with considering the likelihood of seeing a king (or any other card) being drawn at random from a pack, could we not simply say that in a 'pack' of physical systems, all having the same initial wave function f, there is before measurement a proportion equal to the modulus squared of c_j of systems whose physical quantity A just *has* the value a_j? It will be argued that since here too we are 'drawing at random' the system to be measured, it is not surprising that the probability of 'coming up with' the value a_j is then equal to that proportion.

It is quite natural that such a reply should come to mind. But in the conceptual context being considered here, where there are no 'extra variables', it cannot be sustained for the following reason. Let us *suppose* that before measurement the 'pack' of systems we are looking at was indeed made up of parts in each of which A had a fixed value. If we then decided to measure, not A but another quantity B, and, applying the appropriate rules to each of the parts separately and combining the results, we tried to calculate *with this assumption* the overall probability that the measurement of B made on any element of the 'pack' comes up with a specific result, we generally would get a wrong answer. This is because the value in question is in general different from that obtained by a *direct* calculation carried out starting with function f and applying the same rules. In general therefore, we cannot assume that in quantum systems the physical quantities have values prior to being measured, a situation very different from that obtaining in classical physics.[8]

However, even though it is difficult, after all, to translate particular statements of quantum physics into the language of strong objectivity, we must not allow that fact to persuade us *without further ado* that it is totally and utterly impossible. We need to look at the matter more closely. Indeed, the question is often posed by physicists and philosophers: if we were to use an appropriate degree of abstraction, would it not after all be possible to consider the mathematical formulation of quantum physics as being *in itself* a sequence of statements expressed exclusively in terms of strong objectivity? To do so, would it not be sufficient to consider that alongside the usual physical realities described by numbers, vectors and functions there are others, just as 'real' and 'physical', described by things such as 'operators'? Those who entertain this conjecture refer quite often to a technical reformulation of quantum mechanics that specialists call 'Heisenberg's representation'. Formally it resembles classical mechanics (which as we know is objective in the strong sense), the only difference being that the numerical values depending on time which describe the motion of bodies in the latter are replaced by *operators* depending on time. The idea therefore seems quite simple: let us show how open-minded we are and accept, alongside realities in themselves that can be described by numerical values, the existence of realities in themselves that can be described by operators. This will give us a quantum physics formulated in strongly objective terms.

Here again, because it is qualitative and general, the suggestion is at first quite attractive. But once more too, we need to examine it closely. To avoid going twice over the same analytical ground, the simplest procedure is to make use of the proven and uncontested fact that Heisenberg's representation is mathematically equivalent to the representation of quantum mechanics we have implicitly used so far (called 'Schrödinger's representation' and expounded in most elementary texts on quantum mechanics). The effect of this equivalence is that the difficulties we have met already in our theoretical study of measurement operations *re-emerge in more or less identical terms in the study of the same phenomenon in the Heisenberg representation.* More generally, the advanced technical research of scientists trying to move beyond narrowly phenomenalist positions to study in detail the problem raised by measurement has led practically all of them to the view that the business is not simply a matter of being open-minded. In their view, within the 'orthodox' formalism the problem necessitates at very least a detailed investigation that must be based on hypotheses concerning the instruments used, on thermodynamic considerations, or on developments of some other kind.

Under the circumstances, there are two possible general directions of research still open to us. One we have just alluded to, and its main aspects will form the subject matter of the next section. The other, which is bolder – or perhaps more foolhardy – is an attempt to change what we have called the 'orthodox' formalism. Obviously, this has to be quite specific. For example, simply stating that a reduction of the wave packet violating Schrödinger's equation occurs objectively 'in certain cases' without in any way physically specifying what these cases are, would be about as pointless as the practice dear to the doctors in Molière of explaining that opium is a sedative because it has a sedative virtue. (We might note in passing that, paradoxically, an over-exclusive taste for the abstract formulations of quantum axiomatics may promote this tendency in certain minds.) On the other hand, as soon as an honest attempt is made to be in physical terms as specific as possible, it becomes clear that this is an extraordinarily difficult undertaking, given all the predictions of the formalism as it currently stands that must be reproduced, for they have all been empirically verified. We should also note that when we think about it, an undertaking of this kind appears rather strange. In science, the great theoretical upheavals have so far been caused by internal necessities: new data need to be explained, the coherence of old formalisms needs to be restored. We must go back a long way into the past to find a theoretical upheaval caused solely by demands of a philosophical kind such as realism makes.

So to come back to the question at issue in this section: we must clearly recognise that whatever the value mathematical abstraction has proved to have in the most varied fields, this does not by itself alone seem likely to provide the key to the problem that exercises us. It is true that theoretical physics has been led to make ever-increasing use of mathematical formulations of its principles, formulations which are very abstract and hence of very wide application. This may well foster the illusion that the ideal of 'mathematical realism' is getting nearer and perhaps has even been achieved. A closer look leads to a very different conclusion. What we are in fact approaching when we proceed thus is simply the ideal of a synthetic representation of phenomena. In other words, it is the ideal of a description of *communicable human experience*, in a form at once concise and universal, and *not* that of a description of reality in itself. This is a point to which we shall return.

6 'Macro-objective' attempts

When one 'microscopic' system (a particle say, to focus our thoughts) interacts with another physical system (a measuring instrument say) it is a fact that after the interaction the composite system is described in general by a wave function that takes the form of quantum superposition. And as we saw in the last chapter, this superposition simultaneously contains states corresponding to macroscopically different physical situations. One false notion must be ruled out immediately, however. *In practice*, if the second system is a measuring instrument it is not possible through further experiment to reveal the fine detail (minute interferences and the like) that rules out any interpretation of the superposition in purely classical terms expressible in ordinary everyday language. Briefly, we could say that such details, although measurable in principle, are in practice too fine to be accessible to the observer. Under these conditions, it is clear that there is no reason why it should not be possible, *within the framework of weak objectivity*, to construct a coherent physical theory of the measuring operation. Indeed, it is not hard to find in the specialised literature articles and books describing just such theories. Their authors quite rightly point out that their theories lead to the same results irrespective of the personality of the experimenter carrying out the measurement, or the sequence of compatible measurements when several are involved, and so on. They generally add that they are concerned only with the results of measurements *actually recorded by the instrument*, for one cannot legitimately speak of unrecorded results. With regard to the words 'recorded' and 'instrument', their definitions are not always identical, but every effort is made to ensure that their own is unambiguous, in the sense that when applied their definitions ensure that any human being can be sure whether or not this or that process is a 'recording by means of an instrument'. In short, they demonstrate rigorously that their theories actually meet the requirements of *weak objectivity*, though they may not use that particular expression (on this see also the Addendum to this book).

We, however, are concerned with a different question. For if we are analysing the conditions under which quantum theory is compatible with physical realism, clearly what we must demand is *strong objectivity*. Hence we cannot – in particular – be satisfied with a *weakly* objective statement of the quantum axiom regarding the result of a measurement. Nor, it should be noted, can we accept 'immunisation' along the lines that, 'with regard to theories of measurement, weak and strong objectivity merge because the "reality" they describe is precisely *the process of measurement*'. It must be stressed that the question we are tackling is of so general a nature that it goes beyond the study of that particular process. Rather, it is none other than this: is the hypothesis of the universal and absolute accuracy of the predictions of quantum formalism compatible with the demand for a strongly objective formulation of the laws governing reality?

Now that we have clearly set out our aim we can examine in some detail the theoretical analyses. These we can arrange under two headings: those which postulate that there are no extra parameters; and those which allow that there might be.

Analyses of the first kind have been developed much more fully than those of the second. Their very diversity shows that in this area it is far from easy to come up with a theory likely to satisfy all specialists. The reasons these specialists give for their dissatisfaction with this or that theory (which often leads them to produce their own) are of course specific and so very diverse also. In general however, they may, in the language we have been using, be described as observing that with regard to some particular point the theory they comment on critically is objective only in the weak sense. . . .

There may be theories of measurement operations which totally escape that kind of objection, but I must say that personally I do not know of any (see for example [2] [27], and the Addendum). Of course, this does not prove there are none, or that none might appear in future. But even if they did there would still be certain stumbling-blocks. I am thinking in particular of what seems to me the impossibility of reconciling, within a theory that is objective in the strong sense, the 'reduction of the state vector' occurring during measurements (or similar interactions), with relativity.

We need to be quite specific here, for once more we are touching on an area in which all kinds of misunderstanding are possible. Within a theory which is objective in the *weak* sense of the term – such as the quantum theory of fields, for in the usual mathematical presentation its basic elements are the operations of preparing systems and measuring observables – there is *no* truly fundamental difficulty in reconciling quantum principles and relativity. (True, there are certain difficulties in the formalism and in calculations, such as the appearance of infinite sums, but specialists have come up with practical ways of satisfactorily getting round such obstacles.) This point needs making once again: where objectivity is weak – which is perfectly satisfactory as long as we are concerned with science in the strict sense of the term – there is *no* inconsistency in contemporary mathematical physics. To take what follows as implying that there is would be a mistake.

Even so, it is nonetheless true that within a theory aiming at *strong objectivity* there is a real conflict between relativity and the idea of the reduction of the state vector interpreted as a *fact of reality*. This has been demonstrated convincingly, notably by Aharanov and Albert [28], [29] in highly technical articles too complex to consider in detail here. (Suffice it to say that the arguments developed by Aharanov are generally found sensible by the international community of mathematical physicists, a body which is usually extremely and justifiably cautious about giving its approval.) Specifically, these authors give a negative answer to the question of whether *a relativistic quantum system can be described in terms of a certain subset of its physical quantities (by giving the precise values of these quantities) even at moments when the system is not being observed directly.*

Their conclusion is that in the relativistic case, quantum states defined by the values of (compatible) physical quantities making up such subsets quite simply have no independent existence.[9] Here, it will be noted, we are dealing with what in our terminology is a counterfactual definition, since it refers to points in time 'when the system is not being observed'. So what is demonstrated is the impossibility, in the relativistic case, of a counterfactual definition of quantum states of systems. This clearly establishes that such quantum states cannot be seen as realities in any non-relative sense, for as we have seen, the notion of physical systems that are real in such a sense – the fundamental notion of the philosophy we have called physical realism – has a place in our conceptual baggage only through the play of counterfactuality. Therefore, in particular, we cannot see the quantum systems in question as attaining in this sense any sort of 'real quantum state' after a measurement operation (or after an interaction with systems with properties similar to those of an instrument). Consequently we must recognise that, as already stated, the reduction of the wave function when a measurement is being made cannot be interpreted as constituting a *real fact*.

Can Aharanov and Albert's reasoning be extended to cover cases where there is no hypothesis to the effect that hidden parameters do not exist? For technical reasons I believe that it cannot (see [2A]). However it would be most questionable to use that observation as the basis for an attempt to reconcile relativity and quantum physics within the framework of a strongly objective theory. As we saw earlier in our discussion of determinism, the obstacle to that undertaking reappears in a different form in each of the many and varied models which admit hidden parameters, that it has been possible to construct. What makes this more especially plain is the difficulty of making real use of the distinctive characteristics of these models in order to account *in detail* for certain correlations at a distance between results of measurements, despite the fact that such characteristics (influences travelling faster than the speed of light, for example) have been introduced specifically in connection with this problem.

So we must close this chapter on a rather negative note. It is of course true that properly speaking there is no *impossibility theorem* that would conclusively establish the non-existence of any strongly-objective interpretation of microphysical phenomena. From this point of view, the situation is less clear-cut than it is with regard to the principle of separability, since at least in principle, as we have seen, the latter can be refuted with all due rigour. Nevertheless, during the period of more than half a century in which microphysics has had a *de facto* existence, and has made more and more advances in the most varied fields, none of the attempts to formulate it in terms of strong objectivity has proved able to surmount all the difficulties the problem entails sufficiently convincingly to compel acceptance. That at least is a matter of objective fact, and in our search for a vision of the world that does not leave physics out of account, we simply cannot ignore it.

7 Physical realism in general: a conclusion

As we defined it in Chapter 1, physical realism is nothing other than the thesis that the term 'independent reality' (by implication, 'independent of human minds') is not only meaningful, but designates an entity that in principle is knowable by means of science, even if at the present moment it is not fully known. To accept this thesis clearly means that we must regard such knowledge as the goal of science. It therefore also means we must consider that knowledge of the practical rules of prediction, even though it may, as has been stressed, constitute genuine knowledge, is clearly not synonymous with the kind of knowledge that is our ultimate goal.

Because scientific knowledge has a kind of certainty – at least a relative one – which distinguishes it from conjecture pure and simple, anyone who deems the thesis of physical realism to be well-founded should expect physics to produce increasingly general theories and should also expect these not to be enduringly in conflict with one another. In each field there should therefore remain just one such general theory, once the short-lived period of trial and error is over, and it should be possible to formulate these general theories as descriptions of reality. This latter condition can also be expressed by saying that the general theories in question must be capable of being stated in terms of strong objectivity. Now if we look at the way (which is universally accepted, the few variants not affecting its substance) in which quantum mechanics is usually formulated, we can see that it meets the condition of uniqueness but not that of strong objectivity. If on the other hand we consider the various theoretical attempts we have reviewed here, we note that those among them which have amounted to something (non-local hidden variables theory, relativity of states theory, and we might add the 'implicit-order model' [31] that we did not mention) do meet the condition of strong objectivity; but we also note that there is more than one of them and that we have no means of deciding among them scientifically.

From this, it seems to me, we must conclude that physical realism is an 'ideal' from which we remain distant. Indeed, a comparison with conditions that ruled in the past suggests that we are a great deal more distant from it than our predecessors thought they were a century ago.

Part III

Causality, reality and time

7
Irreversibility

Like art and literature, systematic thought has its specialist disciplines; and as in art and literature, to change from one to another tends to be disconcerting. Instinctively we prefer continuity, within a single conceptual and linguistic framework implicitly defined once and for all. If we were to seek the underlying reasons for this attitude, we should probably find them in the fact that, in the early stages of the acquisition of knowledge, serious investigation was successful only when it was confined to carefully circumscribed area, corresponding to just one of the specialist disciplines (mathematics, philosophy, psychology, and so on).

By no means am I saying that we have now left that stage entirely behind us. To do so would open the way to all kinds of sloppy thinking, for it is clear that switching disciplines carelessly and uncritically leads to glibness of the worst kind. And yet there is no doubt that systematic thought is gravely hampered by an ideal of purity that obliges every specialist, be he a mathematician, a physicist, a biologist, or a philosopher, to contemplate reality solely through the spectacles of his own specialism. Human knowledge is now so extensive that to reflect on the 'great questions' in this restrictive way is to deny ourselves the possibility of a balanced view. As far as this book is concerned, this means that from now on we shall have to pay ever more attention to matters of fact and of principle which lie increasingly further outside the strict framework of mathematical physics and which consequently require us to switch disciplines to investigate them. But these switches should not prove too disconcerting, and in any case will be made as rarely as possible and in full appreciation of the difficulty of performing them satisfactorily. In the present chapter they are barely noticeable, since the questions we must now address require only very slight excursions into other fields.

These questions are two-fold. In the first place, towards the end of the last chapter we saw that fairly serious philosophical difficulties arise (at least for 'realists') if we set the hypothesis of the universal validity of the prediction rules of quantum mechanics alongside the actually-observed facts. We shall have to consider whether these difficulties can be overcome by using the converse hypothesis, namely, that such rules are pure idealisations which are valid only in particular fields. Second, there is at the moment an interesting line of research directed towards the detailed investigation of the relations between the laws of elementary particles on the one hand and macroscopic phenomena of the other, notably in respect of non-equilibrium phenomena and the development of complexity. We shall need to consider to what extent the results of such investigations complete (or perhaps modify) those provided by the study of the microscopic world. Although these two questions are obviously closely linked, not to distinguish between them, or to do so incorrectly, leads to confusion and perhaps serious errors of judgement. So here we shall consider them separately. There is some advantage in starting with the second and – to begin with – it is appropriate to remind ourselves of both the *problem* and the *sense of disquiet* brought about by a certain *distance* that has gradually become apparent between the macroscopic world and the world of elementary phenomena.

1 The question of distance

Like all other human beings, scientists have an undeniable tendency to underestimate the time, energy and resources needed for the success of any new project. This optimism (which to be sure plays an important part in getting projects under way) was in former times reflected in the view of many physicists that the only really important goal is the discovery of fundamental laws which, once known, would make the study of complex phenomena merely a matter of computation, if not of routine. The recent immense progress in computer calculations seems at first glance to bear out such a view. Rather surprisingly however, it also coincides with a new awareness within the scientific community of our relative inability actually to reduce the behaviour of complex systems as they are found in nature to the operation of fundamental laws alone.

To illustrate this phenomenon, some physicists have used the metaphorical but expressive notion of saying that the equations used to describe such elementary laws are *more intelligent than we are ourselves*. Others add that it seems necessary to *complete* such equations by some qualitative principle, for example one having the effect of selecting a time arrow. On this last point, which is clearly very important, there is unfortunately no general agreement, and the discussions have been neither sufficiently conclusive nor focussed enough to be summarised in a few pages. Generally speaking, we should however not imagine that physicists have to deal with a complete separation between the 'elementary' and the macroscopic. In a sense, it is going too far to talk as we did above of our 'inability', even if we tone down the expression by adding 'relative'. Nor could there be anything more wrongheaded than the notion that there are two *separate* physics, based on different principles. The fact of the matter is that far from being unbreachable, the wall separating the elementary from the macroscopic has recently been breached in several particularly interesting areas – such as turbulence phenomena, dissipative structures and the like – through developments that only required very plausible hypotheses.

It must also be pointed out that, however – and this is the main point I wish to make here – such developments are extremely arduous, call for an abstract and subtle mathematical apparatus, and have so far produced only very limited results. Indeed, the distance between the elementary and the macroscopic, especially macroscopic configurations out of equilibrium, has been shown to be considerable.

2 The phenomenon of incomprehensibility

There is, then, in a quite objective sense a 'distance' between the microscopic and the macroscopic. Accompanying this objective fact there is, unsurprisingly, a kind of subjective disquiet, excellently highlighted by Prigogine [32]. We are faced, it may be said, with a 'new incomprehensibility'. We can see what is meant by this if we refer to previous relationships between knowledge and descriptions of the Universe. In the eighteenth and nineteenth centuries for example, astronomy provided those who knew something of it, even if they were not specialists, with a genuine enlargement of their vision of the world, a bounty which had the three-fold advantage of being at once authentic, inspiring and encouraging. It was authentic because of its undeniably objective nature, inspiring in that it revealed an immense spectacle, and encouraging in its proof of the power of reason. And that proof was clear. The 'thinking reed' could understand the world, something that the successful prediction of the return of Halley's comet made plain to all. During those centuries, the primary emphasis was still on the *enduring nature* of the great motions of the Universe, as opposed to the apparent fidgetiness of everyday phenomena. The fact that the laws of such motions were comprehensible was therefore also a contributory factor in producing a sense of deep agreement between the grand phenomena such as tides and eclipses, and the most exact scientific theory. The result was a well-justified sense of harmony.

Although that was and still is the feeling produced by the indisputable fact that there are such things as universal laws, it has to be admitted that it is counterbalanced by the sense, more clearly perceived with each passing decade, of a dissonance (I avoid the word 'discord') in the field of evolutionary phenomena. This dissonance, like the harmony just described, is perceived to exist between the observation of large-scale phenomena, such as the evolution of living things, of the stars, and of the Universe itself, on the one hand, and on the other the very nature of the fundamental laws.

To grasp the essence of this division, we must first note that in recent centuries an awareness of the irreversibility of events, be they physical, biological or historical, has been an increasingly important aspect of the thought of human societies. It is true that for over two millenia the idea of cyclical time, most valued by the Greeks and Hindus, has been more or less explicitly opposed by the ideas of some major monotheistic religions, which are centred on an historical view of time. However, this clash is in no way an absolute one, as indicated by the adjective 'eternal' that such religions are not reluctant to ascribe to God. It became more substantial when the increasing pace at which life-styles were changing forced societies to become more acutely aware of the changes taking place within their own sphere.

Although the concept of time as used in classical physics readily lends itself to the description of stationary phenomena such as those of traditional astronomy, it is not so readily adapted to serve in even the merely qualitative description of the wealth of essentially evolutionary phenomena. It is far too closely 'akin' to space for that. This feature of time is brought out in the reversible nature of the basic equations of physics. Schematically, this consists in the fact that if a sequence of events satisfying these equations were filmed, an observer subsequently watching the film would have no way of deciding whether the film was being projected forwards or backwards. If the movement being calculated is, for example, that of a satellite of Jupiter or Saturn around its mother planet, such 'undecidability' seems to accord quite well with what actually happens. However, if we try using these equations to predict quite simple developmental phenomena such as the growth of a plant or, more simply, motions that involve considerable friction, this undecidability seems to contradict our most reliable experience.

To be sure, the contradiction has been eliminated, at least in a sense, by the theoretical developments of statistical mechanics. But that does not eliminate the intuitive *dissonance* we have encountered. This is all the more so, since until very recently statistical mechanics seemed effective only in the description of phenomena tending towards 'uniformity in chaos', that is to say, towards what is often called 'thermal death'. The 'open' phenomena that at least temporarily correspond to an increase in the degree of orderliness or the building of more complex structures do not in any way contradict statistical mechanics, but they do seem to fall more or less clearly outside the domain in which it is able to make predictions. But do not such phenomena underlie the whole world of our actual experience, and hence everything that is of greatest importance to us? It is true that the development of various disciplines which have taken their place alongside physics but remain outside the scope of its laws has often made it possible to account for such phenomena, at least in qualitative terms. But it is also true that this has not dispelled the unease. Nowadays it is a matter of fact that people of sound judgement who may not be well-versed in microphysics but who do meet experts in that field often tend to feel that however knowledgeable these experts may be, *something essential seems nevertheless to have eluded them*.

3 Suggestions for new ways forward

Some present-day theorists and thermodynamics specialists believe that this essential something can be at least partially grasped. They base their belief primarily on the fact that during the last decade or so there have been important theoretical advances which at last make it possible, using the elementary laws of physics, to understand in qualitative terms the processes by which complex structures may be formed. Of course, the step from the idea of 'catastrophies' or 'dissipative structures' (to mention just two of these recent developments) to a quantitative and truly general understanding of how living things develop is still quite considerable. There is also a long way to go before we reach any agreement about what the basis of such an understanding will ultimately be. But there has undeniably been some progress. It is to be hoped that there will be more, and that in times to come we shall see the gradual emergence of a science which, far from reducing the *élan vital* to some humdrum mechanism, will reveal to us all the more clearly its profound beauty, much as classical astronomy once revealed the beauty of the Universe to our forefathers.

For some physicists however, such simple and, of course, still excessively vague hopes are not enough. They want to set out the way rather more precisely. The task that faces them necessarily involves two stages, which – it must be appreciated – call for developments in opposite directions. The first is to surmount, in one way or another, the problem of the reversibility of elementary physical equations.[1] In particular, they must account for 'plain' irreversibility, as manifested in the evolution towards thermodynamic equilibrium. The second task (note that it is only second) is to explain how certain 'open' systems are characterised by the opposite development, namely an increase in their internal complexity and a diversification of their structure.

It is of course legitimate to hope that algorithms can be devised which will be applicable to both aspects of irreversibility, inverse but complementary as they are. That is why we need to dwell for a moment on the first of these aspects, even though progress made in understanding it has been to some extent public property for a while now.

Indeed, it is well known that the early steps in the study of this first type of irreversibility were taken a long time ago, and a great deal of effort has since been devoted to the subject. This is not the place for a detailed history of the problem, referring to such matters as Boltzmann's 'H theorem' or the objections raised by Loschmidt and Zermelo. Nor can we try to summarise the classical discussions of such problems (which are in any case to be found in numerous excellent treatises on statistical mechanics). The one thing we need to bear in mind, for it will prove useful, is that although such problems are raised by both quantum and classical physics, for obvious historical reasons it is only within the framework of the latter that they were first investigated and various conclusions reached. In general, it proved possible to transpose these to the quantum domain, but it was often easier to deal with them within classical physics. This explains something that the layman might otherwise find surprising, namely that although the laws of classical physics are incorrect when applied to 'microscopic' objects such as elementary particles, atoms and molecules, many investigations of the type we are considering are in fact conducted within the framework provided by those laws. The real justification for such a procedure is that the investigation of any highly complex problem must be carried out in stages, and that solving problems of the kind considered here within the axiomatics of classical physics has in the past often been a very effective stage in devising the most fruitful approaches, even though the ultimate aim was a solution based on the principles of *contemporary* physics and hence one that referred to quantum rather than classical fundamental laws.

As has been stressed more than once above, the fundamental laws of classical physics can be formulated in terms of strong objectivity. Indeed, this the easiest way of stating them. Consequently, classical physics was instinctively seen, both by most of those who developed it and most of those who commented on it, as falling within the scope of the philosophy we have called physical realism. In other words, notwithstanding a few dissenting voices, it was generally deemed to provide a good description of an independent reality, rather than a description of merely the collective and communicable experience human beings can have of such a reality. Within this conceptual framework however, the usual explanations of thermodynamical irreversibility – meaning those provided by statistical mechanics as currently taught – cannot but create some philosophical difficulties. The reason is that these explanations are based on the hypothetical homogeneity of small-scale probabilities ('coarse graining'), a hypothesis that is readily validated if we accept arguments based on human inability to carry out certain measurements (such as those of the individual positions and velocities of 10^{23} molecules), but practically impossible to justify if we seek to eliminate *all* reference to the capacities and potentialities of human beings. Both students and teachers of statistical mechanics are fully aware of this inherent difficulty in their subject; and in practice many of the latter make meritorious pedagogical efforts in trying to reduce it, or more precisely, to 'sweep it under the carpet', as the familiar phrase has it. As with the conceptual problems of quantum mechanics, this difficulty *in no way* affects the predictive ability of statistical mechanics. Hence those who teach the subject quite rightly tend to minimise this particular problem. Nevertheless it is also true that outside the lecture room the physical realists among them are on the whole somewhat disturbed by it . . . .

It is at this level that the new schools of thought intervene. 'There can be no question' – they say – 'of accepting that macroscopic irreversibility is merely an illusion, or of attributing it to our ignorance.' And this kind of protest is seen by many primarily as a refusal to accept a conceptual dichotomy: the one that arises, as stressed above, from the fact that on the one hand classical physics makes possible and indeed favours the philosophical position in which fundamental laws are said to describe an independent reality (reality 'as such'), while on the other hand once this position has been adopted the same classical physics allows us to understand such a manifest phenomenon as macroscopic irreversibility only as an 'appearance for human beings' (even if it demonstrates that this appearance holds *for all of us*).

Obviously, however, those who raise this objection do so only as a stimulus to new lines of investigation. As scientists, they are quite aware that a flat refusal is no kind of argument; but it can, nonetheless, prove fruitful if it leads to further research. In the present case, a refusal of such a kind acted as an incentive for numerous outstanding research projects, of which that of the Brussels School (associated with I. Prigogine) is among the best known. This research is based on a fact highlighted by recent studies of microscopic/macroscopic relations already referred to. Certain macroscopic systems in a state of non-equilibrium are so unstable that *even in classical physics* the notion of precise simultaneous values for the individual positions and velocities of all the molecules composing them (briefly, for all the *dynamical variables*) loses all scientific meaning. This means that even in the very unlikely event of there being two physical systems of that kind which at some time have the values of all their dynamical variables *as nearly identical as we please*, these values would generally be appreciably different shortly afterwards. In such circumstances, knowing the initial state of each molecule in a system (supposing our knowledge were quasi-perfect, which it cannot be) would not enable us to predict how that system would behave subsequently. To do that we should need *absolute* precision, which as we all know, is a purely theoretical notion. So from the point of view of scientific practicalities at least, we must give up the idea of dealing with a *specific* system and the precise values of its dynamical variables, and instead look at an *ensemble* (statistical set) of systems and the statistical distributions of the values of the dynamical variables within that ensemble. Such distributions are in fact the only entities about which statistical mechanics enables us to make verifiable predictions.

Some of the new schools of thought mentioned above see this as indicating that we need to take quite a decisive epistemological step. For example Prigogine [33] considers the question of whether or not the state of affairs just described is purely practical in nature, and says he is 'rather inclined to believe that it is of considerable theoretical and *conceptual* significance [my italics], since it forces us to go beyond a purely dynamical description'. In the context of this quotation, the phrase 'a purely dynamical description' means a 'description which takes as its objects of reference the dynamical variables of the various molecules composing a *single* system', and the claim that we must *go beyond* such a description must be taken in a strong sense here. It amounts to saying that, at least from a scientific point of view, such dynamical descriptions *are not meaningful*.

Later we shall reflect on the philosophical significance of such a step. Here we shall note briefly how it forms the starting point from which the new schools of thought go on to discover irreversibility at the very heart of a physics which can in a sense still be called microphysics, since the equations in question, far from violating the fundamental laws, are to some degree an extension of them. At the stage we have now reached, having ruled out descriptions by means of dynamical variables (or to put it another way, descriptions which refer to an *individual* macroscopic system), we find that we have at our disposal equations which describe the development of only an ensemble of such systems, or more precisely, of the 'phase space density'[2] (conventionally designated by ρ) relative to this ensemble. These equations, however, are still reversible. If we think about it, there is nothing surprising about this, given that the equations in question are always – by a simple conceptual about-turn – interpretable in terms of the dynamical trajectories of individual systems. Even if we decide that such a step is not necessary the mere fact that it *could* be taken is enough to explain the reversibility we are discussing. This means that if we are to attain irreversibility we have to go one step further. Along with other scientists the members of the Brussels School suggest that this step must consist of a mathematical transformation of density ρ (also called a 'distribution function') into a new distribution function $\tilde{\rho}$. As in all transformations from one function to another, this one can be represented by what mathematicians call an 'operator'. Now what is significant here, as has recently been demonstrated, is that certain types of unstable systems are such that this operator can be chosen so that the new function $\tilde{\rho}$ undergoes an irreversible time evolution. Correspondingly, it is possible to define a 'microscopic entropy' operator for these systems. As Prigogine points out [33], once this operator is introduced what becomes fundamental is the description in terms of distribution functions, and it is no longer possible to reduce the distribution to individual trajectories. Given what has been said, we can see that this last point seems decisive.

If it is really the description in terms of distribution functions – more especially, the description in terms of $\tilde{\rho}$ – that is fundamental, if it is truly *no longer possible to reduce these functions to dynamical trajectories*, then we can see that the whole world view created by classical physics is shaken and perhaps shattered. A certain *reductionist* approach, all too often implicit in popularising works and adopted – inadvertently to be sure, or with a view to trying to 'simplify matters' – by too many genuine physicists when writing books aimed at a wide readership, stands condemned. Thus for example texts in which the early stages of the Universe are described in terms of thermal agitation of particles in collision, but with no indication that such language is purely and simply allegorical, are unacceptable, even when written by most eminent physicists.

Much more interesting is the fact that under these conditions and given the irreversibility of the development in time of $\tilde{\rho}$, the *irreversible* phenomena studied in nineteenth and twentieth-century physics (including those which generate order and complexity, and are of the greatest interest to us here) acquire once more a 'status' at least as 'grand' as that of reversible phenomena. (What, beyond our immediate intuitions, we should understand by this will be discussed later in terms of 'reality'.) Indeed, within the formalism devised by the Brussels School a notion of systems out of equilibrium having an *age* has emerged. This 'age' is very different from normal mathematical time. Unlike the latter, it is formally described not by a numerical variable but – as with all observables in quantum mechanics – by an operator defined in a functional space. A further similarity with quantum mechanics is that the 'age' operator has a specific value only for some, not all, the states of the system. It is much better adapted than mathematical time to describing changes affecting open systems out of equilibrium, particularly those which manifest tendencies towards greater order and complexity. In other words, this formalistic 'age' resembles the age of our immediate experience more than normal mathematical time does, and thus comes closer to representing the duration we experience as living systems.

4 Discussion

These brief summaries of recent developments contain some important points for scientists and some significant ones for philosophers. The second, however, do not totally coincide with the first. For the scientist reflecting on these things *qua* scientist, the second stage of the process – the transition from ρ to $\tilde{\rho}$ – is the more interesting and therefore requires the greatest attention, all the more so since it is based on subtle and sophisticated mathematical techniques whose play and modes of application offer a fascinating spectacle. For the philosopher – or for the scientist wearing his philosopher's hat – the first is the more important. As a matter of fact, to conclude that description by means of dynamical variables is not 'scientifically fruitful' in the case of interest to us (and must therefore be abandoned by scientists in search of fresh results) does not *of itself* lead to the conclusion that in such cases the dynamical variables are meaningless. It does so only if we make a philosophical choice, which is quite legitimate and can be supported by arguments, but which nevertheless must still be clearly perceived *as* a choice.

This choice is to adopt instrumentalism as a ruling principle, or at least as the only methodology acceptable to the 'scientist'. Indeed, we should bear in mind that initially neither ρ nor $\tilde{\rho}$ has to do with an *individual* (macroscopic) system. These functions were introduced from the study of *ensembles* of systems. So far, no author has suggested considering them as data bearing on the *independent reality* of *individual* macroscopic classical systems, knowledge of which would exhaust any conceivable descriptions of such systems. In fact, it is impossible to see how an interpetation of that kind could be sustained in any coherent way. Abandoning dynamical trajectories thus means giving up the idea of an independent reality that, in principle, could be described. This is why Prigogine, when he proposes defending the idea of going beyond the limits of dynamical descriptions [33], quite rightly stresses the arguments that are in favour of the choice mentioned above. 'We are' – he writes – 'always tempted to describe the physical world as if we were not part of it. If this were the case . . . the initial conditions could be stated with infinite precision. But such an external view of the world *is not the object of physics* [my italics]. Our task is to describe the physical world through our measurements and to remember that we are part of it. . . . One of the main objectives of theoretical physics is just that of making explicit the general limitations introduced by measuring procedures.' Indeed, these few words could in a way be seen as constituting a plea for us to conceive physics as centred on weak objectivity, or to put it another way, as a reasoned declaration of faith in instrumentalism. And it is clear that he *had* to make such a declaration. It is obvious that anyone who does the opposite and adopts a strictly Spinoza-like view of classical physics, seeing or wanting to see in it a description of an *intelligible independent reality* (or at least of some of its features), could not be convinced of the 'reality' of irreversibility through reading about the advances in mathematical physics stemming from the Brussels School. For such a person, the irreversible behaviour of $\tilde{\rho}$ would always be an *appearance*, a merely subjective reality which, seen from his own viewpoint, would be purely phenomenal and hence of secondary interest. (In the Addendum the reader may however find a basis for discussing the possibility of a middle course between pure Spinozist realism and strict instrumentalism).

It would obscure the sense of the fine work mentioned above to see in it simply a paradigm shift from a physics of 'Being' to a physics of 'becoming' *within an unchanged realist way of seeing things*. Such an interpretation would have the two-fold defect of being inconsistent and of incorrectly representing the ideas of the authors concerned. To be sure, these authors themselves do here and there mention the idea of a shift from a physics of Being to one of becoming, and a rapid reading of what they wrote could therefore give the idea that such a shift is effected by them within the framework of a theory that meets the demands of physical realism.

However, even a marginally closer reading is enough to convince us that this is not the case.

Nevertheless, becoming aware of this cannot in itself be a sufficient conclusion to the sequence of questions that spring to mind in this connection. In Prigogine's observations, a believer in physical realism, strong objectivity and pure Spinozist thought – or simply a layman uninitiated in all these debates – might easily find elements which are of the nature of mere assertions (and which he could then accuse of being 'gratuitous') rather than of the nature of obvious and compelling truths. While this author states, for example, that the objective of physics is not to take 'an external view of the world', it can safely be said that a disciple of the mature Einstein would take exactly the opposite view. This being so, it seems fairly clear that the true 'objective of physics' cannot be posited *a priori*, or at least that the decision to do so could not meet with unanimous approval. The same holds good for the statement that the task of explaining the general limitations introduced by measuring processes is one of the fundamental objectives of theoretical physics. That statement too is not a self-evident truth.

Moreover, the author quoted himself acknowledges that it is not, since he writes in the same place that just that lesson is what first relativity and then quantum mechanics teach us. But is it *really* the *incontrovertible* lesson of both these theories? It seems doubtful that it is a lesson relativity would impose on us, since, again, the creator of that theory – Einstein – himself finally adopted physical realism, and hence a completely opposite point of view. In fact, as we have seen, relativity (particularly general relativity) nowhere contradicts such realism. Indeed, what it asserts, essentially, is that we *take views* of independent reality, views which depend on our own situation as well as on reality itself. This is an important insight, but it does not in any way contradict the idea of an intelligible 'reality in itself'. Indeed, in this connection it is easy to think of an analogy which though no doubt rather summary is not unreasonable to mention here: the description of a vector by means of coordinates. These relate to a frame of reference; changing the frame of reference clearly implies changing the coordinates. But that does not mean that the vector itself depends on the frame of reference.[3]

Then again, is the instrumentalists view of things a lesson imposed on us as a result of quantum mechanics? We have seen that it is not, or more precisely, not if the word 'imposed' is taken in its strongest sense and if by 'quantum mechanics' we mean only the conventional formalism. Indeed, prior to the fairly recent work referred to in Chapters 5 and 6, there were ways of getting round the problem that might seem attractive to advocates of physical realism, notably the hypothesis of local hidden parameters and the postulate of the 'determinism of the wave function'.

Thus at least as long as we forget about non-separability, the situation is that there are no truly decisive arguments in favour of the Brussels theses. To illustrate this, let us suppose for example that *classical* physics holds good even at the microscopic level (which is not the case, but as we said it is the convention adopted as an heuristic hypothesis in much of the work outlined above); or let us assume that the hypothesis of local hidden parameters is acceptable; or again let us assume that the notion of quantum states of individual microsystems is generally acceptable, with 'states' understood in a realistic way. In each of these cases, the relevance of the part of the Brussels work that relates to the basic irreversibility problem could not easily be defended against objections raised by the supporters of so-called 'naive' realism. (Of course we leave aside all other parts of the Brussels work, which do not raise any special philosophical problems and the pertinence of which is quite beyond discussion.) Indeed, no convincing argument would then be at our disposal against either physical realism in general or even the particular case of atomistic 'multitudinism'; and we all have so deeply rooted in us the idea of an independent reality composed of localised objects that it would then be natural to picture the world as being composed of such micro-objects existing in themselves and obeying the laws of microphysics. Now, whether classical or quantum, these laws are in any case reversible. . . . Again, the irreversible development of $\tilde{\rho}$ would, on such an hypothesis, seem to be merely an intersubjective view we take, one which consequently does not deal with the essence of things.

However, what must fundamentally change our attitude, and dispel most of our reservations, is precisely the fact that all these hypotheses are purely academic. Classical physics does *not* hold good at the microscopic level, the hypothesis of local hidden parameters *cannot* be defended, and the quantum state of particles *cannot* be objectivised generally and in the full sense of the word. In other words, the apparently negative conclusions we reached in Chapter 6 play a positive role here. It is only thanks to them that we can finally pass a positive judgement on the theses discussed above. Indeed, without in any way adopting instrumentalism as an *a priori* tool in our investigations, we have been able to see *a posteriori*, after a methodical examination of contemporary physics *as it actually is*, that it is virtually impossible to express that physics in the language of strong objectivity. And it is only because of this finding that we can, at the present advanced stage of our discussion, give the philosophical ideas of the Brussels School the importance they deserve.

5 Questions of vocabulary

Does it then follow that we must accept all the explicit or implicit meanings that school of thought gives to the words it uses? Such questions of language are more important than they seem, and merit some examination. The particularly sensitive areas are those connected with the terms 'ignorance', 'illusion' and 'reality'.

As we have seen,the Brussels School's view is that earlier explanations of irreversibility were based on our *ignorance* of fine detail, in other words that irreversibility was seen as an *illusion*, and that one of the School's most important achievements is to have shown the consistency of a different way of seeing things, one in which ignorance does not play such a disconcerting role and irreversibility is a *reality* and no longer a mere illusion. We have also seen, however, that it would be a mistake to understand this claim as simply saying that the work in question describes a reality which is both *totally independent of human beings* and *irreversible*. The reality it deals with is one that we cannot describe as if we were not part of it. In other words, it is a 'reality for us'. We shall frequently use the expressions 'empirical reality' or 'sensible reality' to refer to such a reality.

This being so, the first question that arises is whether or not it makes sense to talk of a deeper reality. Is the idea of independent reality, or reality in itself, meaningful? The question is very difficult, since it forces us to reflect on the meaning of the word 'meaning'. We shall not try to avoid doing so, but we can put off the task for the moment, for it is clear that the physicists of the Brussels School raise the idea of such a reality only in order to highlight the contrast between it and their own conceptions; in fact they reject it. Provisionally, and within the framework of the present discussion, we must therefore play by the same rules and give the word 'reality' just the meaning mentioned above, namely that of *empirical reality* or *reality for us*.

However, this raises a further question connected with the meaning of 'illusion'. Illusion is that which is not real. The traditional derivations of irreversibility refer, as we know, to certain limits, intersubjective but nevertheless inescapable, in our observational capacities. If the word 'real' has to do only with reality for us, then such limits are 'real', or at very least it is not obvious that they are any less 'real' than a $\tilde{\rho}$ function. In both cases, there is reference to the same notion of intersubjectivity. If we accept the conception of reality adopted by the Brussels School, the grounds for regarding irreversibility explained in the traditional way as 'illusory', and the behaviour of the $\tilde{\rho}$ function as 'real', seem – once we think about it – much less clear than one might at first have supposed.

We shall not prolong the discussion of this particular point, since (contrary perhaps to the opinion of the authors considered here) it seems to be of secondary importance. Ultimately what is important is simply that there is a sense of word 'reality' in which irreversibility – and in particular the constructive irreversibility of, for example, the *élan vital* – is not an illusion; that this sense, particularised by the adjectives 'empirical' or 'sensible', is the only one which has direct contact with communicable experience; and that this is true not only of everyday experience but also of scientific experimentation, including that in the microscopic field.

6 Are the quantum rules merely idealisations?

This is the second of the two questions we set out to examine. Some authors, notably those whose ideas we have just considered, like to stress that in effect both classical and quantum mechanics are mere *idealisations*. Such a claim is interesting, even attractive in some respects, since it expresses a thesis that has a certain rationality. We must admit, for example, that considering our present stage of knowledge Newtonian mechanics now appears as but an idealisation of reality based on the fact that in the investigations of a number of phenomena (such as that, from among many others, of the motion of the planets) it is possible to pay no attention to some processes, which though they are important in other fields are not in the one in question. It is conceivable that the prediction rules of quantum mechanics might one day also look like that to those who come after us. In other words, they might turn out to be relevant only in those particular fields in which they currently seem to us useful and indispensable.

Considered in itself, the hypothesis is admissible but clearly unproductive. It needs the addition of some positive element which – and this is the difficulty – cannot be pure fantasy. It must be constructed on a solid and specific basis. The trouble then is: if both quantum and classical analytical mechanics are viewed as tainted with relativism and reduced to the level of mere models, one valid only in the microscopic domain and the other only in definite macroscopic cases, where is such a basis to be found? There is obviously some danger of a vicious circle here.

We may wonder whether the theoretical speculations we have been following in this chapter offer some way of avoiding this danger. It is far from obvious that they do, for they are based from the very start on the principles of classical or quantum mechanics. It is by building on one or the other of these systems of principles[4] that, through a process of refinement and purification, they construct a physics that can be applied to phenomena which, considered from the point of view of these two kinds of mechanics, seem too complex to be studied effectively. Somewhat audacious, then, may the ideas of some of their proponents appear when they conceive of turning, once such a physics has been constructed, against classical and quantum mechanics, to accuse them of being mere idealisations. If neither kind of mechanics describes for us the true laws which the constituent parts of macroscopic systems in some way obey, if both are *no more than* idealisations (rather like the notions of the vault of heaven, the ecliptic, and so on) it is hard to see how they could have formed a good starting point for the construction of any physics of these same macroscopic systems. Our surprise at this would be of the same kind we would experience if we discovered that modern astronomy had been built up by a process of refining and purifying that ancient example of 'idealisation', the Ptolemaic theory of epicycles.

A tentative response could of course be that the road to discovery leads through byways and thickets, and that although every idea must subsequently be tested by reasoned procedures, the stages by which it originally comes into being do not have to follow a strict chain of reasoning. But the difficulty still remains. It is moreover appropriate to bear in mind that at the beginning of the present chapter we justified the question we are considering by referring to the philosophical difficulties previously encountered while examining the hypothesis that the rules of quantum mechanics are universal. But such difficulties (non-separability and the like) become major problems only, as we have seen, within the framework of the philosophy we have called 'physical realism', that is, within the confines of a philosophy centred on the contention that the concept of an intelligible and describable independent reality (situated in space-time) is meaningful. Since, as we have also seen, such a philosophy is rejected out of hand by the authors whose investigations we have examined here, the difficulties in question can scarcely have seemed acute to them. Bearing this in mind, it is hard to see how to base *on their work* the idea that quantum rules – or principles – are no more than 'idealisations'.

We cannot therefore rule out the possibility that the idea in question is but the expression in their language of the idea that I express in my own by the notions of *weak objectivity* and *purely empirical* reality. If this is the case, the claim that quantum rules are merely an idealisation simply means that although in principle these rules are universally valid there are areas – important areas, in fact – of our scientific experience in which they are not to be used for two reasons, one of them being that in practice the rules in question are inapplicable given the complexity of the physical systems involved, the second being that other rules (those of classical mechanics, the thermodynamic rules governing systems out of equilibrium, and so on) can be used easily and with success. If this exegesis is deemed acceptable, I can withdraw the objection set out above, since acknowledging the fact that, in principle, quantum rules (leaving aside the classical ones) are universally – and exactly – valid means that reference can be made to them to the full extent to which they cast useful light, despite the possible complexity of the problems at hand.

The fact must not be hidden that such a way of seeing things, which I endorse, supposes that the idea expressed by the words 'in principle' is not an empty one. In other words, it implies that the rules we have been discussing, unlike the model of the vault of heaven, bear some structural analogy to what really *is*. (To this point, the most difficult of all we have to deal with in this book, we shall return later.) It is nevertheless true that, for our present purpose, such a structural analogy is not necessarily equivalent to actual knowledge. Consequently, it can be seen that the solution proposed here does not amount to a return to the thesis of *physical* realism (or if the term seems preferable, to that of the existence of an *intelligible* independent reality). It is a *rational* response to the difficulties we have encountered, but one which differs from positivism or classical instrumentalism as much as it does from – at the other end of the spectrum – traditional Spinozist thought.

8
Sensible reality

In quantum physics the notion of measurement is fundamental (it is present in the axioms of the theory), and there seem to be no credible measurement theories that are truly objective in the strong sense of the word. Hence present-day theoretical physics is objective only in a weak sense.

There seems to be no gainsaying this fact. Nevertheless, we would attain only a simplistic view of knowledge if we did not try to understand to what extent that weak objectivity is still genuinely objective.

1 Sensible reality is real

As a guide on our intellectual journey, we can use the reflections of physicists such as Asher Peres [34]. Peres examines the interaction between a system being measured and the instrument used for this purpose. And he asks what new measurements would in principle have to be made if we wanted to observe a difference between two ensembles E and E', defined as follows. Their components are systems of the type called S in Chapter 6, section 4 (a composite system consisting of a system and a measuring device with which it has interacted). In E however, all the S are supposed to be in one and the same *quantum superposition* (as defined earlier), whereas in E' every S, and in particular its 'instrument' part, is in a quantum state corresponding to a *specific* value (indicated by the device) of the measured quantity, the number of systems S having one specified value being proportional to the corresponding observational probability as yielded by the formalism for the case being considered. In principle, there is generally a difference between E and E', and there is no fundamental principle of physics from which it would be possible to infer that this difference is unobservable. Peres, however, shows that in order to observe it, we would have to be capable of measuring, on the instrument itself, a physical quantity that *in a certain sense* cannot be measured. In trying to say exactly what he means by this, he admits that in the case in which the instrument is endowed with only a finite number of degrees of freedom, it would not be theoretically 'impossible' to measure the relevant physical quantity. However he also points out that in the case of macroscopic instruments the price would be something like a non-negligible increase in the entropy of the Universe. Is it conceivable that human beings, even at a more advanced technological stage, might one day be able to pay such a price? Given a sufficiently large number of degrees of freedom in the instrument – and in what is connected to it – the answer is certainly no. Indeed, if we tried to assess the feasibility of the operation it would quickly become apparent to us that possible 'cosmological limits' are involved. The time such an operation would require might be greater than the lifespan of our planet. This is brought home all the more forcefully by the fact that, as Wigner [35] and Zeh [36] have shown, at the level of sophistication involved here there is no place, not even in interstellar space, where a macroscopic system can be considered to be isolated.

The objectivity of such limitations, it must be stressed, is merely *weak*, in the sense we have agreed. Such limitations have explicitly to do with the possibilities of operations carried out by human beings in their current situation, inhabiting a planet one day doomed to annihilation (and in a Universe that is perhaps also destined to disappear, though that is no doubt more of a moot point). Claiming to see such limitations as endowed with strong objectivity would mean claiming they can be expressed in terms relating to physical reality in itself, quite independently of the spatio-temporal circumstances obtaining within the Solar System, our galaxy, or even larger-scale systems; there seems no way of achieving such a reformulation. We can, however, see that the objectivity of such limitations, though weak in the 'technical' sense of the term, is nonetheless very strong in the sense in which we instinctively tend to understand the epithet. As a systematic study would show, this last observation could be made generally valid. The objectivity of modern physics is weak in the technical sense, but it is very strong in the essentially intuitive sense that we tend to give to the word.

This observation is of considerable importance, since it allows, or rather obliges, us to take seriously the idea of a *sensible reality* (or *world*, if the term is preferred) already referred to on more than one occasion earlier. The fact of the matter is that empirical reality – or the sensible world, or sensible reality – is simply the totality of what we immediately experience, either directly or through the medium of our telescopes, microscopes and other tools. Among other things, it includes the world of objects, the 'things' (*res*) from which its name is derived. It also includes the whole *evolution* of both living creatures and the stars or the Universe. We used to think we could elevate most of its components, including those that incorporate data on positions or velocities, into independent realities, things 'as such' or 'in themselves'. We now see that a mental operation of such a kind is no longer possible. Considered *sub specie aeternitatis*, that concept of empirical reality is essentially anthropocentric.

Remark

Other arguments of a less scientific but more eloquent nature spring readily to mind in support of such a judgement. If for example there were a society of intelligent computers, what kind of physics would they construct? When we think about the question, we soon see that there are a number of important factors to be taken into consideration before we answer it. Without claiming to exhaust the list, we may note that on the one hand computers easily distinguish between numbers, whereas on the other the idea of qualitative forms and differences between forms is not readily accessible to them. For us as human beings, the opposite is the case. Two objects of the same shape lying in the same direction at distances from us that are numerically different only in the eighth decimal place are considered by us virtually indistinguishable. For the computer they would be radically different. Conversely, two objects of different shape are always distinct for us even if they are in practically the same place. For the computer, the difference between them might at a first approximation be negligible.

Since approximations are necessary in any kind of physics (to be able to talk of distinct physical systems without involving the whole of the Universe, we must, as an approximation, treat systems which are not quite separate as if they were so) it is clear that the physics of a society of computers would be essentially different from our own. Even the way they would describe to each other their everyday experience of reality would be fundamentally different to ours. There seems no way round this, even if to simplify matters we suppose that the 'objective events' behind the two kinds of experience are identical.

2 Three pitfalls

The question is how are we to apprehend this sensible reality? To answer it properly we have to be on our guard against three pitfalls.

The first is to forget that we are in fact dealing with *sensible* reality, not 'matter' as imagined by some kind of naively-reductionist nineteenth-century science. The infinite subtlety which sensible reality unfolds before our eyes is *not* a childish illusion masking a robust elementarity; the traditional scientist who simply muttered 'carbon, oxygen, hydrogen, nitrogen' when he encountered a tender glance or gazed upon a majestic forest deceived himself. Not everyone finds it easy to fight this particular prejudice, and even when he is aware of it the man of science – or the man trained in scientific disciplines – often has a particularly hard task in this respect. But he has to make the effort, if in his daily round he wants to experience fully a spontaneous freshness which is in fact nearer to truth than the reduction to the quantitative and the elementary that the burden of his scientific knowledge would force upon him if he let it. At this point we may remember Marcel Proust's aphorism that we feel in one world and think and name in another. It applies to all of us, but characterises most sharply the anguish of the scientist. And when Proust goes on to say that we can establish a concordance but not fill the gap between these worlds, those of us who belong to the scientific community feel even more in agreement than those who are strangers to that community. To reduce the spiritual impoverishment created by such a divide there is no better way than constantly to bear in mind that sensible reality can in no way be reduced to the few elementary concepts of a science that would be *elementary in its very concepts*.

The second – opposite – pitfall is that of reducing to the everyday level the whole of what we consider to be meaningful. Men and women of action are probably those most threatened by the dangers this represents, but curiously enough even pure scientific researchers can get entangled in it. How many of them, obliged as they are to be *in practice* matchless technicians in leading disciplines, eventually become no more than technicians *in spirit*, with no other goal than technical success? We could go further. In their case, paradoxically, the risks relative to the two hitherto considered pitfalls combine and reinforce one another. A scientist caught up in and absorbed by the admirable power of the particular techniques of his own field, and having therefore no time for critical analysis of its guiding principles, is more than anyone else – particularly if he is not himself a specialist in theoretical physics – inclined to picture science (in which he masters but a small sector) as a sort of rigid ladder resting on a theoretical physics conceived of – again – as conceptually elementary, and as the sole ultimate source of truth.

The third pitfall is the systematic rejection of science, a rejection one might be tempted to make by what has just been said. It *is* a danger, for what science has to tell us about empirical reality is obviously very important. Refusing to listen to it inevitably leads to mere invention or the adoption of crude and obscure mythologies. Astrology, chiromancy and the like are chief among the former, while this or that philosophical 'system' which sweep aside the pedestrian methods of the scientist exemplify the latter. With regard to our picture of the world, completely abandoning the guiding thread and the safeguards provided by science generally means falling a prey to charlatans.

3 Understanding the sensible

Having seen where the pitfalls lie, we can resume our journey. Once again: how are we to apprehend the sensible reality it is science's vocation to describe? By science alone? By science supported by philosophy? If so, by what kind of philosophy? A few preliminary remarks will guide us in answering such questions.

The first, very often heard, is that unlike science philosophy is not based on experimentation. This is a very real difference between the two disciplines (even though it needs qualifying), but it is not of course the only one. Another is that the mode of reasoning of certain philosophers is fundamentally different from that of scientists, and less susceptible to guidance by strict criteria. On the other hand, there is one matter (to which we must pay attention here) about which the scientist is most often *less* demanding than the philosopher. It is a question, as the reader has probably guessed, of the foundations of basic concepts. *From the very beginning* philosophers have reflected on the nature of time, space, matter and the like ('What is time?' St Augustine asked long ago. 'If no-one asks me, I know. If I have to explain it, I cannot'). By contrast, scientists tend to take such questions seriously only in carefully-defined, specific circumstances; more precisely, they are willing to give them a little of their attention only when forced to do so by theoretical and experimental developments. It is in 'good faith' and with no qualms whatsoever that the physicist used to deal with macroscopic objects only – at least if he is not an astrophysicist – writes *t* in his equations and considers that by doing so he has disposed quite simply of all the meaningful problems the nature of time might conceivably raise for him. In the same way, for the philosopher the idea of an object must necessarily be grounded on some piece of serious thinking before it can validly be used (hence the *a priori* notion that objects can truly exist only in relation to a subject). For the scientist, on the other hand, that idea is not at all a compelling necessity *ab initio*. Quite the opposite in fact, since he is naturally quite ready to adopt in this case, as he did with the idea of time, an intuitive idealisation of the concept of an object such as that we effect in everyday life. It is only when constrained to do so – by the discoveries of twentieth-century physics – that he accepts the need to distance himself to some extent from such extremely simple intuitions and to recognise that in certain cases the very concepts of time or of an object constitute in themselves significant problems.

Even so, we still have to make some distinctions. With regard to time for example, the scientific community has admitted, since the advent of relativity theory, that the concept the post-Newtonian scientist had of it must be amended, and it now accepts the view that, in principle at least, our new conception of time applies to every part of physics (even if in some of them the older one is still useful in practice). As for the concept of an object, however, it has on the whole (there are exceptions) adopted a viewpoint some would describe as more sophisticated and others less consistent. With regard to those objects we have conventionally referred to as 'microscopic' – particles, atoms and small molecules, for which the prediction rules of quantum physics are the only ones that are operative and practically applicable – the scientific community is perfectly willing to admit, and to proclaim, that the common intuition associated with the 'concept of an object' has to be abandoned. On the other hand, with regard to 'bigger' objects of which the former rules of 'classical' physics hold good, that same community, imbued with the spirit described above, adopts a much more conservative attitude, generally deeming it appropriate to retain for them the 'concept of an object' bequeathed by traditional common experience. In their view, to wonder about the validity or non-validity of this intuition in such cases is to run the risk of getting lost in airy-fairy speculation, and all to no purpose.

Such observations may smack a little of social psychology but nevertheless they have real cognitive significance, especially if we relate them to our discussions in the first section of this chapter. Taken together, the facts we have noted explain why the scientific community tends to confuse – and in a sense *quite rightly* – in its usual way of expressing its ideas strong and weak objectivity, on the one hand, and independent and empirical *reality* on the other. Though it can legitimately be seen only as weak, 'objectivity' in the normal scientific sense is nonetheless a true objectivity. Even if it is no more than empirical – or sensible – the reality described by science is just as real as that of birth and death.

One very significant consequence of this has to do with what we should make of a statement like the following: 'human intelligence, which has been capable of discovering and hence understanding evolution, *is itself merely a product of evolution*'. Upon reading such a statement, we might at first be inclined to consider it as incompatible with what has been said in this book. In particular, if we are aware of the fact that evolution – irreversibility – is purely and simply a phenomenon, that in other words it relates only to a *human* reality, more precisely to a reality that is a reality only for the human intelligence, we might feel that to claim that such intelligence is itself a product of evolution is incoherent. If – we might protest – all empirical reality is ultimately no more than a picture composed by the human mind, then human beings and human intelligence naturally precede and are causative of the picture in question, and so cannot be effects of it. Taking this argument into account we might well be tempted to say that the statement 'intelligence is the product of evolution' makes sense only within a philosophy of physical realism, that is, within a framework that data provided by present-day physics has already led us to reject.

Such reasoning, it seems to me, neglects certain important facts and is therefore not satisfactory as it stands. The cause/effect dichotomy is too crude to be adaptable to much of the data provided by even the most positivistic kind of scientific research. *A fortiori* we must be careful not to let it shackle our ideas in a field as far removed from everyday experience as that of the questions we are discussing here. There seems to be no logical way of disproving the thesis that between mind and sensible reality there is a complementarity relationship which gives them their actuality, and the claim we are discussing (that intelligence is the product of evolution) can always be defended, even against its most incisive critics, as an admittedly partial and certainly metaphorical but in no way *false* expression of that complementarity. Physics is of course full of examples of 'simple models' which, although not to be taken literally, nevertheless contain nuggets of genuine truth. Whenever a scientist, anxious to keep both feet on the ground and using a way of thinking that, after all, is appropriate to his aims, draws no distinction between 'reality as such' and the empirical reality constituted by 'what is seen to happen' on our planet (and by extension in the Universe), such a scientist is correlatively justified in considering our intelligence as no more than a product of evolution. Once again, to say that is about as objectively true as to say that we are born and die. In our examination of the question of sensible reality we shall henceforth adopt that point of view.

However trite this may seem, it clearly has important implications. Evolution is based on the struggle for survival (a view which has been generally accepted since Darwin's day; it would serve no purpose to digress here to discuss the justification for it). Accordingly, human intelligence, like the shells of insects or the claws of carnivorous animals, must be seen as a tool to serve in that struggle. The fact that it proved so useful means that those individuals or groups most highly endowed with it were automatically selected for survival. But is the obvious usefulness of *knowing* equally great in relation to *everything* that could, conceivably, be known? Here certain nineteenth-century authors, whose answer was 'no', seem to me to have had the right idea. During the ages when the human species was coming into being (and still to a certain extent today) *technical expertise* was incomparably more useful than abstract knowledge. Not that such knowledge – or pseudo-knowledge – served no purpose. Divination, magic and great myths have certainly played a vital part both in promoting group cohesion and in providing a source of much-needed encouragement and consolation in adversity. However, the basic knowledge without which the epiphenomenon of the abstract could never have arisen, was nevertheless that which had to do with hunting, making fire and fashioning tools, and more generally every kind of technical skill.

Knowledge of this kind, which could indeed serve as a tool in the struggle for survival, had to do essentially with objects. The hundreds of millenia during which the human brain was formed thus formed it in such a way that far from being a transcendental instrument of absolute knowledge (a kind of knowledge which, with rather obvious inconsistency, some evolutionist materialists believe to exist), our intelligence is only a partly-effective tool, providing knowledge of admittedly quite a high level of generality but chiefly directed to what lies *outside* ourselves and our species. Is it, due to this genetic inheritance, centred on solids? As already noted by some philosophers, we might well believe so, when we observe the contrast between the relative ease with which our intelligence has mastered the geometry and mechanics of 'point masses', as in classical astronomy, and the effort it had to make to develop classical hydrodynamics, and now has to make on a much grander scale to gain understanding of the mysteries of complex systems that are out of equilibrium. And what of the effort called for when our intelligence turns to the problem of the living world? It is not that it fails in these endeavours. Progress is indeed made, albeit through sheer tenacity. But even limited progress in that field comes considerably harder than complete solutions, expressible in mathematical terms, in the sciences of 'matter'. The secrets of the *élan vital* – the undeniable propensity of life to become increasingly complex – are no longer as inaccessible to our understanding at the end of the twentieth century as they seemed to be at its beginning. Here, we are making progress, albeit with great difficulty. Soon, thanks to science, we may come up with a fully quantitative explanation of *some* aspects of the question, perhaps. . . .

Can we ever reckon to explain all of them? If we take into account all the foregoing considerations, it looks highly unlikely. It is true that – by taking explicitly into consideration the merely *weak* nature of the objectivity of the theory they used – Prigogine and his collaborators were able, as we have seen, to get back, by scientific means, to the notion of *age* (of systems out of equilibrium), an idea very similiar to the one that Bergson arrived at intuitively and called 'duration'. It is also true (the observation is important, and we shall return to it shortly) that Bergson's advocacy of duration as a rival to 'real time' and his philosophical approach of *basing this notion on the very awareness we have of it* [37] have been at least partly rehabilitated by the similarly weak objectivity of any theory of empirical reality, whereas to the old physical realism such an approach could only appear nonsensical. However, between that and the contention that issues such as duration and the *élan vital* can better be grasped *in all their aspects* by purely scientific methods than by any other means there lies a large gulf. Considering the various points we have been discussing, such a contention would indeed constitute quite a rash judgement.

However, there are those who would retort that there is no other way of apprehending such entities, or at least none that can be deemed valid. This point of view calls for some discussion. It involves *adopting a position* supported by a *semantic convention*. To be sure, if we start from the position that by definition the only valid ways of 'grasping', 'apprehending', or 'knowing' are by means of the discursive intelligence, the channels of which are, it seems, mainly structured by the left hemisphere of the brain, the conclusion follows logically. It follows, however, essentially from a linguistic convention which, like all conventions, can scarcely be based on anything other than reasons of convenience. Are such reasons enough to justify fully the very restrictive decision that we cannot consider any of the information mainly prepared by the 'right hemisphere'[1] to be genuine knowledge? In my view they are not.

Rather, I take quite the opposite view and consider that through, for example, the notion of *age* as devised by Prigogine 'right-hemisphere knowledge' acquires a kind of scientific respectability. I base this claim on three linked reasons. In the first place, the notion of age is scientifically respectable because it is part of a range of genuinely scientific developments (dissipative structures and the like). Secondly, the only way we have of avoiding seeing the 'age' in question as a mere 'appearance for us' (in contrast to mathematical time – or space-time – seen as an element of independent reality) is to give up any claim to strong objectivity, that is, to abandon hope of an authentic description of independent reality (either we assert that it is not precisely knowable or we reject the very concept). Indeed, as we have seen, Prigogine's way of reasoning is to posit that what is inaccessible in practice (such as determinacy in a system with a 'strange attractor') need not be considered – is not 'real'. When applied to space and time, once all due precautions have been taken, such a procedure amounts to generalising the Bergsonian approach described above for duration, and basing space and time on the (collective) *consciousness* of space and time. In this way, it is possible to introduce without inconsistency the idea that *age*, or Bergsonian duration, is in fact more truly real (in the empirical sense) than the reversible classical time of the majority of physicists, since the latter (a) has turned out to be but an element of a recipe and (b) is never the object of our immediate intuition. Third and finally, it cannot be said that this *a posteriori* scientific justification of the Bergsonian notion of duration is merely a Pyrrhic victory for Bergson, and the graveyard of his first idea (centred, as we have seen, on the inability of human intelligence alone to grasp the essence of duration and movement, and hence on the need, in this matter, for some recourse to sensibility, or in his words, to intuition). Indeed, although scientific intelligence can now, after considerable effort, assimilate *some* features of the notion of duration and is therefore no longer obliged to judge such a notion totally foreign to its own ways of seeing things (and this is a most important advance), it is also true that, obviously, it is by means of an activity linked to sensibility, to emotion, that we grasp the most significant aspects of duration most directly. Prigogine is quite right to raise the idea of the 're-enchantment of the world' in connection with the work on irreversibility discussed above. It is clear, however, that any such re-enchantment necessarily proceeds via an emotion which is itself largely subsidiary to the global apprehensions, be they spatial or musical, which, it seems, we develop in the main through the right hemisphere of our brain. This means that if the effect of the Brussels analyses were to attribute exclusively to the *left* hemisphere, or more exactly to the analytical thought that hemisphere specialises in, any real apprehension of anything, then any hope of the restoration of enchantment would be a delusion. What gives support to that hope is, again, that such work is not to be seen in that light. Rather, it should be understood as the discovery by the dominant (left) hemisphere that some modes of apprehension which seemed naturally to belong to the 'right' hemisphere only are its concern too; and this to a much greater extent than could have been thought in the age of 'classical' science.

9
Independent reality

Many physicists have as part of their mental furniture an amusing fable composed by one of their number. In *The Character of Physical Law* [38] Richard Feynman illustrates the limits of positivism (or *thorough-going instrumentalism*, to use the expression defined in Chapter 1) by telling a fanciful tale involving Pre-Columbian Indians, though deriving more from cliché than reality. It is well known that, at a certain period, the Maya had developed a highly-elaborate calendar based entirely on a radical empiricism that they, rather in the manner later urged by the Vienna School, elevated to the status of a principle. From this factual basis, Feynman tells, we may go on to think of some more or less mythical Mayans who, using ancient records, found the main elements of the complex numerical relationships that link the recurrence over the centuries of the eclipses of the Sun and Moon, rather as the Swiss teacher Balmer was later to discover the formula which bears his name and has to do with the hydrogen spectrum. Once elevated to general laws, the relationships in question enabled the Mayan scientists – or at least those in the fable – to predict with certainty the date of future eclipses. One day, Feynman goes on, a young Mayan graduate student informed his supervisor that he had constructed a theory. Let us suppose, he suggested, that the celestial phenomena whose movements we are studying are really solid material objects at varying distances from us, and also that one of them is a source of light. We could then *explain* eclipses simply as a matter of the alignment of three objects, one of which is the Earth. Elaborating his idea, the young scientist described to his mentor the rudiments of what we should now regard as the correct theory of eclipses.

The immediate reaction of the latter, however, was simply to smile in amusement. But, moved by the disappointment this caused the young man, he agreed to explain why he could not accept the theory. The young man's attitude, he said, was quite naive and pre-scientific. The fact of the matter was that in the field in which they were working, all the rules of prediction were already available. Taken together, these constituted science and all there is to science. Remember, he went on, only *phenomena* are meaningful. Our rules are valid for everyone and are therefore rigorously objective. Imagining the Sun and Moon to be material as you do is pointlessly to seek *explanations* of laws; in other words, it amounts to stubbornly introducing the old anthropomorphic idea of 'cause' into a field where it only complicates description to no avail. To put matters plain and simply, it is to indulge in *metaphysics*, a quite fruitless and indeed shameful thing to do, as our most eminent philosophers will tell you.

So, in essence, runs Feynman's tale. It provides in agreeable fashion a good illustration of the sense of absurdity experienced by most physicists when faced with the systematic negation of the concept of independent reality and the refusal to take seriously anything other than mere observable regularities.

1 It is unthinkable to deny the validity of the concept of independent reality[1]

I intend to provide arguments supporting this statement, which in fact raises serious questions. For the sake of clarity however, I must digress a little in order to avoid some possible misunderstandings that could arise from the terms to be used. Contrary to a widespread habit, even (or particularly) among philosophers, in what follows I shall make no reference to *etymology*.

Etymology is an interesting discipline. Devoting oneself to it can be fascinating, and it may well be supposed that its charm and air of establishing something substantial indicates an obscure but not illusionary influence of valid archetypes. But when we are trying to establish a meaningful line of argument in a difficult field such as the present one, which in many respects is far removed from traditional and familiar experience, it seems dangerous to trust our fate to such a frail barque. To do so would be to give implicitly, indeed unwittingly, decisive weight to old hierarchies of meaning and values, of which we cannot know whether they constitute elements of eternal truth. This applies notably to the words 'real' and 'reality' (for the most part interchangeable as I use them here), and to the verb 'to exist'.

Since 'reality' comes from *res*, Latin for 'thing' and most often 'solid localised object', if we wished to show punctilious respect for etymology, we should not consider light, for example, as an element of reality. That would be absurd. I shall take exactly the opposite approach, ignoring what the derivation of words may have to tell us. And thus I shall define 'independent reality' (or 'reality as such', or 'the real') but *not* empirical reality as being 'what exists' (in itself or as an attribute).

As for the verb 'to exist', I shall again pay no attention to its derivation. Etymologically, it seems to mean 'to occur outside' or perhaps 'to be distinguished from', in the way that an individual may seem to be lost in a crowd but is in fact quite distinct from it, or a drop may separate itself from a liquid in other respects quite homogeneous. In any case, the etymological sense has long since been diluted and even disappeared in current usage. If a theologian tells us that God exists, it is not in that sense that he uses the word. Rather, he gives it a much more profound meaning, corresponding to that of the verb 'to be' in its fullest sense. Essentially the same is true of a mathematician who says of an equation that its roots exist, or a zoologist who tells us that kangeroos exist in Australia. It is also true of the physicist who informs us that although the drop we were talking about certainly exists, so too does the liquid from which it separated itself. In all these statements, the idea of 'emerging' can be discerned only if we engage in some rather unwarranted mental gymnastics, which are dangerous and not justified by any rational argument (as the last example clearly shows). Such is my case for casting aside etymological constraints, at least as far as these words are concerned.

The reader will perhaps observe that 'to be' has still not been defined. Such a remark is rather immediate, but as with a number of almost reflex reactions provoked by the form of the discourse, it only *seems* to be appropriate. We all know that any dictionary is bound to be circular to some degree, for some words can be defined only by more general ones, which in turn must be defined by even more general ones, and so on. Strictly speaking, there is no end to this process.[2] That is why – and 'realists' of all varieties would readily agree on this – strictly speaking we can neither prove nor even *define* 'realism' (since we cannot fully explain by means of language the meaning of the 'is' in the proposition 'I call real everything that is'). However, it would be sophistry – as well as a logical error – to conclude that realism is therefore not true. We all know quite well that we generally think of *something*. All of us too, not excepting any philosopher (as Descartes can be understood to say[3]) are certain, whether we articulate it or not, that at the precise moment when we are thinking, at least something *exists*, or *is*; namely, thought itself or whatever it proceeds from. And that fact alone is enough to refute the assertion that the very notion of existence – of Being – is meaningless. However subtle our dialectics, we could not, it would seem, be logically consistent in claiming both that we think and that the assertion of something we call 'thought' is incorrect. We must therefore conclude that the word 'existence' – in the sense of Being – is not void of meaning since we have seen that in at least one case this word can be used. Finally, it should be noted that to restrict the use of this word to the labelling of elements which belong to a previously adopted linguistic framework (as we have seen Carnap proposes) does not get rid of the objection raised here. For the choice of that framework and the thinking it presupposes are themselves the work of a subject; to say that his existence in his own eyes is *merely* the result of a convention that he himself has established is simply a vicious circle.[4]

These considerations are sufficient, I believe, to enable us to dismiss those unconsidered objections in which all usage of the words 'reality', 'existence' and their various derivatives not conforming to a strictly operational code are immediately ruled out of court. Still, in a way they do not establish fully convincing grounds for realism. Even its partisans admit that realism cannot be *proved* in the way that a theorem can be proved. Nor can it be demonstrated, in the strictest sense of the word, by direct appeal to common sense. Things become even more complex when, as we have seen happen, we have to admit that independent reality cannot be seen as one and the same thing as the set of all the 'events', nor, more generally, as the set of all the 'phenomena'. Feynman's fable, for example, however much it may tell us about the limitations of pure positivism, cannot be seen as in itself proof of the validity of the realist view of things.

Or so at least is the opinion of a number of contemporary physicists. Their starting point is the principle that although it is legitimate to seek explanations of phenomena, by definition 'to explain' merely means reducing the unknown – the yet unexplored – to the known. That, they point out, was precisely what the young Mayan's theory did, since the concept of macroscopic material objects that constituted its basic element was already at hand and was therefore known. Indeed, the supervisor in the fable was therefore quite wrong (as common sense correctly judges) to condemn the way this theory explained phenomena by positing the existence of certain independently-real material objects. But these physicists stress that things are very different in contemporary theoretical physics. This – they say – is because in the present state of our knowledge (described in earlier chapters) no attempt to *explain* phenomena (including their regularity, and intersubjective agreement about these matters) referring to the structure of an independent reality can validly be made in the form of a description of that reality formulated in terms of *concepts already known*. Any such attempt involves taking an intellectual step that is *only outwardly* similar to the – legitimate – one made by the young Mayan; in fact it is much bolder.[5] Is it also legitimate? This is what the above-mentioned physicists ask.

The answer we give them must be that a step of that kind – that is, the realistic approach in the most general sense of the term – is supported by an extremely powerful and in my view irrefutable argument: that once the opposite approach is given precise form it runs up against all sorts of difficulties, even absurdities. If we bear in mind the conclusion we came to above (that *at least* there is such a thing as what we refer to as a thought), to say that independent reality (independent of human minds, that is to say) *does not exist* or *does not make sense* amounts in the end to saying that what can be said to exist is the thinking (human) being and such a being only, or at least that we can legitimately posit existence of other beings only as secondary to his. And that is a notion I for my part reject absolutely, finding myself in agreement on this point with what common sense naturally tells us. Whatever words we use, to make the very existence (of any thing or entity whatsoever) subordinate to human beings (thesis (a)) seems to me to be in the end a position as unwarranted as the solipsistic claim that I myself am the only thing that really exists. It could even be said that solipsism is preferable to thesis (a) in that the former avoids the difficulty encountered by thesis (a) when explanation of intersubjective agreement is called for (a difficulty which thesis (a) can avoid only by denying any need for explanation, as discussed below). The fact of the matter is that although it is impossible to refute solipsism with any rigour, in any case it is not adopted. *A fortiori*, this indicates that we should reject any 'collective solipsism'(!), that is, any view that only human beings, or, what comes to the same thing in the end, human 'consciousnesses', exists, since from the point of view of logical coherence such a conception is inferior to true solipsism.

There is nothing esoteric, nor properly speaking is there anything original in the foregoing arguments; implicitly, they are present in the ideas of all realists (including materialists). What does need to be pointed out is that they remain valid even if the *independent* reality they refer to is not truly knowable in detail by human beings. Their validity does not depend on any particular hypothesis on that score. This means they retain their force even in the situation in which, as we have seen, a number of contemporary physicists consider that in fact we find ourselves.

2 Discussion

Even if, as it seems, present-day physics puts us in a position to take a radically new look at the problems examined in this chapter, these problems are clearly not new in themselves. Indeed, over the centuries they have turned up over and over again in all sorts of guises, with the result that any reasonably detailed account of the arguments and counter-arguments would fill a large tome. A systematic inspection of them would not be appropriate to the purpose of the present book. Fortunately, there are philosophical textbooks which will enable the reader, should he wish, to gain an objective overview of the developments in this area. However, we cannot avoid the need to look, if rather briskly, at some of the arguments put forward by those who doubt that the concept of an independent reality has any definite meaning. We shall restrict ourselves to a discussion of just *some* of the questions raised in considering the concept, notably those which inevitably arise when we introduce, as implicitly at least we have already done, the idea that independent reality is an *explanation* of phenomena.

In everyday life, force of habit induces us to take many things for granted. But if for some reason our attention is drawn to some phenomenon or other, we tend to look for an explanation of it. To look for an explanation is no other than to look for the cause, or system of causes, of the phenomenon in question.

Thus, at first blush at least, the notion of explanation is reduced to that of cause. But here a problem arises. Not without reason, many philosophers have pointed to the need, in principle, for cause and effect to be of the same nature. To their way of thinking, only causes that are themselves phenomena can legitimately be ascribed to effects that have the status of phenomena.

One of the best arguments for this thesis is that all that is clear about the notion of a causal link is that it expresses certain correlations between successive observational data, that is, between phenomena.[6] Neither scientific experiment nor even that of our inner selves reveals anything else. Seeking to ascribe a cause to the fact that such correlations exist is to ask what is the cause of causality, and to pretend to give meaning to a new causal link that is quite different in kind to the one just considered; besides, one of the terms (cause) is here in principle unknown. Such a procedure, we are told, can have no empirical base and so is unwarranted.

If we begin by reducing the notion of cause to that of regularity, which is the idea behind a conception of causality much in favour nowadays, particularly among empiricists (together with all the details and nuances that philosophers who subscribe to it have elaborated), then the objection is undeniably valid, at least if it is a question of introducing the concept of independent reality to serve as a 'cause' of the regularities observed. We should note however both that the conceptual reduction of 'cause' to 'regularity' is not universally accepted, and that if it *is* accepted then we need to ask even more forcefully whether the notion of *explanation* or *raison d'être* can truly be reduced to the concept of cause understood in this way. If we conclude that it cannot, or that it is not logically necessary for it to be, then the idea of an independent reality conceived as the *explanation* of the regularities of phenomena regains its legitimacy.

A theory of cause as regularity as opposed to the theory of cause as 'implication', a theory of explanation as cause as opposed to one of explanation as reason . . . all this draws us into the centre of essentially philosophical problems that have been debated for a long time. I lay no claim to have solved them, not least because I believe it is impossible to do so convincingly and definitively. What we must bear in mind is that so far they have not been solved.[7] Under the circumstances, I see no compelling reason for a cavalier rejection of the arguments presented in the last section, and so take the position that, as an explanation of the regularity observed among phenomena and of intersubjective agreement about them, the notion of independent reality (whether veiled or not only the study of physics can reveal this *a posteriori*) makes sense. If anyone says to me 'but that is metaphysics' I shall simply reply 'so what?'[8]

3 Perspectives on the problem of freedom

We should be hard pressed to find a philosophical question upon which more ink has been spilt than that of the nature, or even the existence, of what we agree to call freedom. Most philosophers have had something to say about it, and their ideas have been astonishingly diverse. Such diversity bears witness to the great difficulty inherent in the matter.

This being so, to come up with a solution to the problem of freedom is, as may be imagined, not one of the objectives of this book. Confronted with a question so many previous efforts have been unable to settle in any definitive way, it would be not without presumption to claim to provide in the space of a paragraph the answer we have all been waiting for, and nothing of the kind should be expected of the next few pages. Instead, what follows are some observations intended to help explain the links that clearly seem to exist between the questions tackled in earlier sections of this chapter and the one we are now examining. A difficult subject is often best approached from several angles; the new perspective afforded by present-day physics offers if not entirely new arguments at least new reasons for favouring some approaches rather than others, together with the means to refine them.

Before going on to those observations, I should no doubt point out that in the opinion of many qualified experts, the new data provided by neurology, however important it may be (I am thinking for example of the possibility we now have of inducing emotions in people *by chemical means*, and hence of instilling motivations that these people will acknowledge as authentically their own) does not change the facts of the basic problem. In any case, such data does not of itself enable us legitimately to refute any one of the philosophical problems, classical or modern, having to do with the existence or non-existence of free will, even if it does lead us to qualify some of them.

As a further preliminary, we should perhaps rehearse in a few words some approaches to these problems (details can be looked up in standard philosophy textbooks).

One of them, usually associated with the philosophy of Spinoza, consists in accepting absolute determinism while at the same time affirming the existence of free acts, which by definition are then actions determined more by conditions internal to the subject than by external ones. This thesis has some substance to it; many so-called refutations which are based on the supposed 'obviousness' of free will in the non-determinist sense of the term, do not stand up to scrutiny. To some extent, the same goes for many would-be refutations based on arguments deriving from ethics and based on the notion of responsibility. But it is no less true that when confronted with at least *some* of the objections of this second kind (examples can be found easily enough), the determinist is hard put to defend his position. As Ewing [39] points out, the determinist may, contrary to what might initially be thought, manage to leave an important place for ethics and moral responsibility, but he cannot *fully* reconcile his position on the matter with the intuitive ethical conceptions we all share. A purely philosophical analysis hence leads to the conclusion that this first thesis is, after all due consideration, rather hard to defend, a judgement indirectly supported by the findings of modern physics, which as we have seen lead us to reject any strictly deterministic theory not allowing for some non-local influences (which would raise difficulties here).

The second thesis before us is, as one may guess, that of indeterminism, or more precisely, that which establishes a strict link between the notions of free will and of quantum indeterminacy by seeing the latter as making the former conceivable. This thesis has been strongly criticised on the basis of the observation, more or less evident in itself, that the indeterminacy of electrons and other sub-atomic particles is very different from the freedom of human beings, and that consequently trying to link the one to the other seems to be a rather simplistic view of things. In the light of present-day knowledge, however, such criticism cannot be taken at face value. Clearly it is genuinely strong only in conjunction with what for brevity we shall call 'the theory of *separate* sciences (or branches of knowledge)', to be discussed below. This theory, though it still sounds attractive to those of a literary – or indeed philosophical – turn of mind has increasingly met with explicit or implicit reservations of a scientific kind stemming from what we have seen is the near universality of the domain of contemporary physical science.[9] Of course there can be no question of *deducing* human freedom, even less so ethics, or the concept of responsibility, from the indeterminacy of electrons: such fields are too remote from each other for such an idea to be thinkable. However, if it is purely a matter of meeting certain objections (which as we have seen are not without value in the deterministic scheme) then reference to quantum indeterminacy – which, in the domain of phenomena, is an established truth, as we have also seen – seems valid. The sole purpose of doing so, however, is to point out that given our present state of knowledge the efforts made by determinists to reconcile determinism with freedom are directed against a problem which does not arise.

A third thesis worth mentioning here may be described as *dualist*. Curiously enough, it attracts minds as dissimilar as, for example, Popper and Eccles. Essentially, it is based on the metaphysical thesis that there is not one but two or even three distinct 'worlds': the world of physical objects, the world of subjective experiences, such as the mechanisms of thought, and finally the world (which may or may not be identical with the preceding one) constituted by the ensemble of human cultural products, such as utterances, reasoning, mathematics and the like. It would go beyond the confines of our subject to show how this thesis differs (at least for those of its adherents with affinities with Popper) from more traditional dualist theses, such as those, more or less Platonist in inspiration, in which the third world as defined above – that of mathematics and the like – is deemed eternal. The only important point to note is that this thesis, far from conflicting with the second, in fact complements it. Once the ghost of absolute determinism has been laid, this dualist view renders the idea that freedom is a *sui generis* property quite natural.

Finally, there is a fourth thesis, which the historian of ideas would perhaps want to link with *parts* of Kant's work. It is founded on the idea that science is essentially concerned with *phenomena* (or empirical reality), as oppposed to reality in itself. The better to describe these phenomena, the diverse disciplines which comprise the 'exact' sciences each formulate their own concepts, which of course they do in such a way that the resulting concepts serve their own purposes. Among the concepts devised in such a fashion, the idea of freedom has no place. It cannot therefore be used by any of these disciplines, which thus normally regard it as a kind of 'foreign body'. But there is nothing surprising in the fact that freedom has no place in the range of concepts in question: far from being a *product* of the exact sciences, it is of the nature of an inner experience, a stimulus to the quest for knowledge, and so a constitutive part of the *foundation* of science.

This regrettably cursory survey of the various theses and approaches to the problem of freedom is of course not exhaustive. It does however raise the main concerns, and so can serve as a reference framework for the observations that form the substance of this section.

Our task, it will be recalled, is to consider whether the re-examination of traditional ideas that characterise present-day physics can help us to form an opinion about the enigma posed by the notion of freedom.

The first observation arising from our survey is that as long as we do not introduce into the study of the philosophical problem any of the specific information provided by physics, the range of possible solutions remains more or less open. Again there is nothing surprising about that: the same applies in many a field (cosmology for example). One of the most important roles of the exact sciences, as has been stressed on several occasions, is to help us make a reasonable choice when we are faced with several *a priori* plausible but mutually exclusive ideas. Can physics serve in this role for the question at issue?

The response must be qualified, for it would certainly be going too far – for example – to maintain that the findings of physics enable us to eliminate three of the four theses outlined above. In all rigour, such findings do not exclude any of them. But as we have seen, they do make the first seem rather improbable; they give it the appearance of an artificial and ultimately unwarranted idea, for which no justification can be found either in inner human experience or scientific investigation. Consequently, such a thesis can be based only on some principle that is posited *a priori* and which is then defended at all costs.

The same cannot be said of the second and third theses (between which there is no radical opposition). If we deliberately decide to remain solely at the level of phenomena, believing – as do many of our contemporaries, especially in 'advanced' countries – that this is the only 'sensible' attitude to take, then as physicists we can readily conceive of freedom after the fashion of the second thesis, where it is linked with indeterminacy, and perhaps venture to specify that link by appealing to the third, notably to the more frankly dualist aspects of the latter (which were emphasised in our survey for that reason). For myself however, I cannot hide my feeling that such a way of seeing things, interesting as it may be, has its defects. It is not that it has been refuted, nor that it is refutable. Rather, it is that it seems to me to be immunised too artificially against any such refutation for doubts not to emerge. Might it not amount to merely an *ad hoc* theory, by means of which it is easy to solve any problem?

Merely raising such a question does not mean that the answer is a foregone conclusion. Indeed, we can continue to think that, at the level of phenomena, the two or three worlds of Popper, Eccles and others are indeed irreducible to one another, and that – still at that level – therein lies the true key to the problem of free will. But when all things have been taken into account, it is the fourth thesis that seems to get nearest to the heart of the matter. Moreover, the findings of physics as I have reported them in this book offer considerable support for it. In itself, of course, the thesis is philosophical, and as such is quite independent of physics. Again, if we wanted to put the thesis in the context of the history of ideas, it would be easy to find roots for it in Kant's distinction between phenomena and noumena, and related aspects of his work. But if we restricted ourselves to such a field, we would be doing no more than sheltering behind a philosophy to which one or more different ones could be opposed. What modern physics seems to be telling us with some force is that these various philosophies – or better, the various epistemologies, various conceptions of the relationship between knowledge and the real which underlie these philosophies – are not all equally credible. From that point of view, the very important part played in the fundamental physics of our day by statements that are only weakly objective, and the difficulty (of which there are very sound reasons to suspect that it might be insurmountable) of replacing without ambiguity such statements with others that are strongly objective, are strong indications that science – and not only sophisticated science, but also that most elementary science, common knowledge – creates its own concepts with above all a view to effective (and quantitative) description. Because these considerations are, as we have seen, the very basis of the fourth thesis on the nature of freedom, it seems necessary for us to accept that this last thereby acquires a clear edge over the other three as far as plausibility is concerned.

It cannot be claimed, all the same, that the argument set out here is equally as convincing as the one we used above in connection with the problem of whether the verb 'to exist' has a genuine meaning beyond the realm of the purely operational. The two are linked, it is true, since the conception of freedom urged upon us by the fourth thesis involves in effect an inner experience that is prior to all description founded on operationalism and so is not unlike our shared conviction that we exist. But the systematic 'doubter', someone who professes to reject all notions whose validity cannot be demonstrated by discursive argument, could still conjecture whether such an inner experience of freedom is nothing but a mistaken appearance. . . . While, even if we concede to him that this experience is no more than an illusion, it would still be true that the very *existence* of the feeling of freedom, illusory as it may be, would still constitute a valid example of the fact that it seems impossible to do without the notion of an 'existence' not anchored in the operational.

Be that as it may, it is clear that this reservation in no way undermines the statement that constitutes the substance of this section, namely, that what present-day physics has to tell us favours the fourth thesis rather than the other three.[10]

10

The dilemma of modern physics: reality or meaning?[1]

After so many pages of philosophy, those of my readers who are physicists – or at all events those physicists who are like myself – will feel the need to return like Antaeus to Mother Earth and renew their strength.

For a scientist, getting back to earth does not necessarily mean getting back to the concrete (nor, on the contrary, to abstract equations). What it does mean is that every forest is made of trees: it is not enough to content oneself with large abstractions, however essential these may be. More specifically, it means steeping oneself *in the details* of the problems of coherence that can arise within these abstractions, and seeking to resolve them with some meticulousness. There is a very effective (if sometimes rather ponderous and repetitive) method that can sometimes be used to this purpose: comparing two approaches to the major problems that are incompatible with each other. If this method is to be reliable, each approach must have been adopted by a scientist who is at once contemporary (and so having full access to all the information we ourselves have at our disposal), and is also a recognised authority in the scientific community. Suppose these conditions are met. What usually follows is that, at least on first reading, both approaches seem convincing not only to the layman (who is always easily won over) but also to those who have made a systematic study of the problem. The incompatibility of the approaches then comes as a surprise even to the latter. It forces them to take a new and more critical look at their initial assumptions. In this way, such a comparison – when it is possible – can indirectly reveal, even to those who thought themselves well prepared, which ones of the ideas they assumed to be established or obvious are mere prejudices.

Are there two such incompatible approaches to the questions that concern us here, approaches which meet both conditions and which if compared can help us further to refine our ideas? The answer, fortunately, is yes. They are presented in two recent articles, one by John A. Wheeler [40], the other by John S. Bell [41].

1 An outline of Wheeler's article

Entitled 'Bits, quanta, meaning', this article could, in a way, be considered no more than an explication of the essential ideas of the Copenhagen

School. Better though would be to characterise it as a description of one possible way of making those ideas manifestly coherent.

Indeed, if as an example we refer back to some of the observations in Part I of this book, we have to admit that the most important of these ideas have been left by their authors (Bohr, Heisenberg, Jordan, Born, and the others) in a form that is not fully and rigorously purged of all ambiguity, and this as a result of a certain oscillation between a *de facto* operationalism and a declared but undeveloped realism. On a more technical level, we may also note that these basic ideas were soon overshadowed by the development of the quantum formalism itself, as oriented by powerful methods mainly elaborated by J. von Neumann and P.A.M. Dirac. These methods centred on the concept of the microscopic quantum state and tended strongly to lend credence, step by step, to the idea that such states – or, what amounts to the same thing, the mathematical entities representing them, such as wave functions and the like – have a reality in themselves independently of the experimental apparatus. As we know, however, Bohr rejected such an idea. From there came the woolliness we noted, not in the formalism itself but rather in the way it is interpreted. It is this woolliness which in its own way Wheeler's article seeks to dispel.

In fact though the article goes further. It could be said that the extreme point of view it adopts (described below) makes it possible to resolve a conceptual ambiguity that, again, is inherent in the position taken by the Copenhagen School, or at least in the idea of it held by a substantial majority of physicists. These latter often think that the position in question is founded on the claim that 'classical' (or 'macroscopic') objects exist in a primary sense which does not need to be defined. In their view, trying to define the concepts of the existence and properties of these objects is pointless, or at best is a purpose that has nothing to do with physics and that physicists should 'leave to the philosophers'. (It is worth noting in this connection that the scientist of the second half of the twentieth century does not, unfortunately, always distinguish, as his forebears knew how to do, between 'a pointless task' and a task 'best left to philosophers'.) Consequently, they think they are being faithful to the position of the Copenhagen School by taking the concepts in question as the foundation of everything else. Their way of presenting this position is therefore to assert (a) that we cannot speak of quantum objects – nor of their properties – in themselves, and (b) that any valid description of quantum phenomena must ultimately refer to the primary concept of classical objects, more specifically to that of the 'experimental set-up' or the 'observational conditions'. As we have noted, such a way of seeing things is often called 'macro-objectivism'. Making use of that term, we may then say that many theoretical physicists regard the position of the Copenhagen School as a particular case of macro-objectivism.

Unfortunately, although macro-objectivism works very well as a practical scheme, we have seen that nevertheless it runs into serious conceptual difficulties having to do with internal coherence. The most obvious of these is that it seems we cannot avoid seeing macroscopic objects as consisting ultimately of microscopic, or quantum, objects. So the main notion that the Copenhagen School wanted to rule out, that of quantum objects and their properties existing quite independently of the experimental set-up, seems to surface again in a problematic way. Again, macro-objectivism seems to require a fairly clear distinction to be made between classical and quantum objects (or events), whereas the equations and formulae of physics do not in themselves provide any indication of how this distinction is to be drawn. Admittedly, it has often been proposed that at least as far as events are concerned it is irreversibility that should be the criterion, but once again it seems difficult to define irreversibility in terms of strong objectivity (that is, in a way that does not make explicit or implicit reference to limitations that are imposed by the human condition on the practical possibilities of action). Even the interesting reflections of a Prigogine, for example, are as we have seen founded in the last resort on the idea that weak objectivity is sufficient (and that it alone is meaningful). There remains, it is true, the thesis that quantum mechanics would only be an approximate theory. But it has been emphasised that for this thesis there is no support from any observational evidence; and it would be a singular occurrence if a scientific breakthrough came from neither the discovery of new facts nor the development of theory but solely from a demand that one may legitimately call philosophical.

But is it truly necessary to identify the philosophy of the Copenhagen School with macro-objectivism? A reading of some texts, those of Heisenberg for example, raises serious doubts. In *Physics and Philosophy* [42] he tackles these questions more subtly. To begin with, he defines what he calls 'an assertion capable of being rendered objective'. For him, an assertion can be rendered objective if it is possible to say of it that its content does not depend on the conditions under which it can be verified. He then goes on to reject as meaningless any 'metaphysical realism' (this is the expression he uses), that is, any realism that cannot be reduced to either 'dogmatic realism' or 'practical realism', as he calls them. These kinds of realism are themselves defined as conceptions in the first of which all, and in the second most, meaningful assertions pertaining to the material world 'are capable of being rendered objective' in the sense defined above. Now from our point of view, what is most significant about this position is not the distinction between 'practical' and 'dogmatic' realism on which Heisenberg insists, though that is also important. Rather, it is that he thinks it is indispensable to *define* what he understands by objectivity and realism, instead of thinking that they concern primary concepts that do not need precise definition. And a second and even more important point of significance is that he defines the notions in question by referring quite directly to the process of verification. In choosing to define these concepts explicitly and to define them in such a manner, he is in effect establishing *verification* as the sole primary concept, or at least as the concept to which even that of reality must definitely be 'subordinated'. This is a decisive choice, the implications of which seem not to have been grasped by most of those physicists who imagine themselves to be substantially in agreement with the ideas of the Copenhagen School.

John Wheeler, however, avoids this inconsistency (if we accept that, properly understood, it truly is one). In his article, the ultimate reference is meaning. This he takes care to define explicitly, indicating that he subscribes to Föllesdal's definition:

> Meaning is the joint product of all the evidence available to people who communicate.

It is true that in the same article the word 'phenomenon' also plays a key role, but we must not be led astray by the fact that in scientific parlance 'phenomenon' is most often used to designate an event taking place within a reality conceived as constituting the ultimate reference. The remark that 'no unobserved phenomenon is a phenomenon', which recurs often in Wheeler's work, is enough to put us on our guard against understanding the term in this popular way. Wheeler in fact takes the term in its etymological sense as 'that which appears and about which there is agreement among observers of good faith'. In this he is not far from Kant. It is essential to understand clearly that in the article in question these are the only two primary concepts (or that 'meaning' is the only one, if we judge that the concept of 'phenomenon' derives from that of 'meaning'). In particular, we could not understand the content of the article if we thought there was another primary concept, namely 'scientifically knowable reality', which would somehow underlay the development of the argument but which the author would have deemed it better not to draw attention to explicitly. In exhorting us to give up, as the foundation of what exists, any 'physical hardware located out there' and put in its place a 'software of meaning located who knows where', Wheeler takes care to rule out such an interpretation. It is clear, moreover, that he does not rule it out only in relation to who knows what 'atomic' or 'quantum' world; on the contrary, he rejects it quite generally.

It is not hard to see that such a radical position makes it possible to remove many of the difficulties encountered by a more realist approach. We shall now examine how this may be done.

2 The method of partial definitions

The role of counterfactual implication has been stressed earlier. We noted that, in a certain sense, such counterfactuals may be seen as lying at the heart of both common-sense and physical realism (the second being in this respect merely an extension of the former). When, for example, I say that my car 'really exists, both night and day', what do I have in mind other than that when I am in bed at night I could if I wanted (albeit that in general I do not) look out the window and see it standing at the kerb? Similarly, if an experimental physicist chosen at random says that a particle emerging from an accelerator has a certain momentum p, what is it that he usually wants to say if not that *if* we were to set up a measuring device on the (rectilinear) trajectory of the particle (something which most of the time we do *not* do) we should record a momentum p?

But is that what a theoretical physicist steeped in 'the spirit of the Copenhagen School', in all its purity, would also say? As we have seen, a reading of the central texts of representatives of that school cannot but raise serious doubts. As we have also seen, these physicists have attached a great deal of importance, in defining phenomena, to the role played by the experimental conditions. And we should not hastily think that such attention is gratuitous. It is not merely a rhetorical flourish. In fact it is the very moment when, faced with the objections of Einstein, Podolsky and Rosen (EPR for short), the Copenhagen School ideas took their definitive shape that we see, through the pen of Bohr, this role emerging as crucial. Indeed, it was what enabled Bohr to 'save the situation'. The essential point in his response was that (here I follow him more or less *verbatim*) to be able to speak validly of 'real properties' of atoms or particles, 'it is necessary to take into account the experimental conditions that define the possible types of prediction about the system since these conditions are an inherent element in the description of any phenomenon to which the expression "physical reality" can meaningfully be attached'. In this response it is particularly important to note that the words 'experimental conditions' are supplemented by those 'that define the possible types of prediction'. Bohr was fully aware that such a specification was essential. If in defining phenomena all we had to take into account were the experimental conditions having to do with *production* (of the state of the physical system under consideration), the EPR objection would be unanswerable, as will be seen in detail if the original articles are studied [43], [44].

So we must accept that according to the 'spirit of Copenhagen' the experimental conditions defining the possible types of *prediction* are essential. Put another way, the specification of the measuring apparatus whose function is to interact with the system under examination must be taken into consideration before we can even begin to speak about the properties of the system; for this apparatus alone determines which of these properties can meaningfully be considered.

In such a situation, it is clear that the procedure of defining properties by *counterfactual* implication is inapplicable. We cannot speak of measurements that are neither made nor intended to be made. In every instance, we must restrict ourselves to the one that *is* to be carried out, and it is exclusively by this means that we must define the properties. But is this possible, we might wonder? The answer is *yes*, at least in so far as the method of defining the properties of objects known as 'the method of partial definitions' (described in Chapter 3) is deemed valid. Developed by Carnap and recommended by Hempel, this method has its critics too, but it does seem to be the most coherent theoretical account of what true disciples of the Copenhagen School are in fact doing when they speak of quantum systems. According to the method, a system S can have property A only if the experimental apparatus makes it possible to measure A. If it does not, then the statement 'S has property A' is 'worse than false'; it is simply meaningless.

The methods of defining by means of counterfactual implications and by partial definitions both appeal to the notion of an instrument. The role of this notion is less fundamental in the former than in the latter, however: with counterfactual implication there is no obstacle to reifying the properties thus defined, seeing them as attributes of the object itself, considered as a thing-as-such. That is why it accords so well, as has been said, with physical realism as with common-sense realism. The same does not hold good for the method of partial definitions, particularly in its application to microphysics, because it totally precludes any reification of properties. We can therefore see that in the philosophy of the Copenhagen School the notion of a measuring instrument is of over-riding importance. This perhaps explains why certain members of this School seem to refer in their writings to a kind of macro-objectivism which in its turn would implicitly refer to a counterfactuality valid in the macroscopic domain (thereby preventing any infinite regress to some 'super-instrument' measuring the coordinates of the first instrument, and so on). We may, however, consider that in Wheeler's article – which we saw makes no fundamental distinction between micro and macrosystems, and rejects the idea of a 'hardware of external reality' – even the properties of macroscopic objects are in the final analysis defined by the method of partial definitions, which unavoidably is centred on man.

It is now possible to state a first, limited conclusion. Wheeler's point of view, however disconcerting it may be (save perhaps for a few Kantians) cannot be rejected out of hand on technical grounds, since a method exists, that of partial definitions, which enables us to define the properties of systems (and hence to practise physics) without the need to make conceptual appeal to the twin notions of counterfactuality and realism. We can also see a little more clearly what is the source of our feeling that his point of view is disconcerting. It originates in the fact that we are not used to defining the properties (and the existence) of objects in terms of the method of partial definitions. Without being clearly aware that we are doing so, we always conceive them as if they were defined by the method of counterfactuals.

3 Has the 'cat paradox' been resolved?

The best way of grasping how and to what extent a conceptual approach resolves the difficulties – also conceptual – raised by quantum physics is to study the way this approach enables us to get round the central enigma of the discipline. This enigma can be presented as follows.[2] Although it is conceivable that a measurement operation might alter a delicate object, the idea that measuring the coordinates of the Moon could substantially alter them must be dismissed out of hand. So too the idea of being obliged by a formalism to conceive that simply taking note of the position of the needle on an instrument (which amounts to measuring its position, whether by another instrument or with the naked eye) can in some cases appreciably affect that position. But (and this is the substance of Schrodinger's celebrated 'cat paradox') let us suppose that a measuring instrument A is made to interact with a particle P (for the purpose, it goes without saying, of measuring some property of P). Subsequently to this interaction consider the system *A plus P*, first before and second after noting (or having noted) the position of the needle of the dial of A. *After* it is quite clear that this position is well defined (since it was noted). *Before*, by contrast, the formalism, if we take it literally (and in its elementary form), suggests that in the most general case the position is *not* well-defined.[3] How can we then avoid the simple conclusion that 'it is the fact of taking note of the position in question that *gave* this position to the needle'?[4]

This is not the place to go into a detailed account of why I believe that none of the numerous 'quantum theories of measurement' so far devised has resolved this paradox (see the Addendum; the interested reader might also care to consult my *Conceptual Foundations of Quantum Mechanics* [2B] on this point). What is of immediate import here is if – and how – the mutually exclusive conceptions of Wheeler and of Bell succeed in doing so. We shall discuss Bell's conception shortly, but may for the moment note that because he accepts the idea of hidden variables his chances of resolving the difficulty seem *a priori* rather good. Wheeler's though, at first sight, seem by contrast small or non-existent, because he does not accept the idea in question.

Such first impressions are wrong in the latter case, however. If the problem seems insurmountable this is because unwittingly we are once again thinking in terms of a reality defined by the method of counterfactuals (in this case this reality is that of the wave function, or in other words, of the state in which the composite system $A + P$ happens to be). If we are truly willing to take the step the members of the Copenhagen School invite us (albeit a little ambiguously) to take, and which Wheeler, quite unambiguously, tells us we *must* take, that is to say, if we give up once and for all the idea of a knowable reality existing in itself 'out there', then the foundation of the difficulty disappears. There is no longer any 'technical' ground for the claim that before we noted it the needle had no definite position.

The best way of understanding why this is so is to imagine we replace the system $A + P$ after interaction with a simpler system, such as that of a particle with spin ½, and the action of noting the position of the needle by that of *measuring the component* S_z, of the spin of the particle along the z-axis. In the case in which preparing the particle (corresponding in our analogy to the composite process of preparing P *and* letting A and P interact) consists in passing it through a Stern and Gerlach device orientated along the x-axis, the quantum physicists tend to say that just before the measurement under consideration the particle has a well-defined spin component S_x along the x-axis so its S_z component cannot then have a definite value (in normal quantum mechanics, two spin components cannot simultaneously have definite values). But in fact, if wc decide systematically to define properties by means of partial definitions *only*, we cannot even pose the premise of this argument, since the particle in question is not, *ex hypothesi*, associated with any instrument for *measuring* S_x but rather with one for measuring S_z. Be it only for this reason, it is impossible to deduce from the axioms of quantum mechanics any refutation of the 'convention' that before the measurement takes place S_z has a definite (but unknown) value.

Clearly this line of reasoning can easily be transposed to the real problem under investigation here. We must therefore acknowledge the correctness of the claim made above, namely that it is only the fact that the human mind does not take naturally to the method of partial definitions but rather always tends implicitly to think in terms of an external reality defined by the method of counterfactuals, that throws into sharp relief the difficulty we encountered here. In fact, the difficulty is real only with regard to the above-mentioned elementary account of quantum mechanics, in which the state before measurement is counterfactually defined. Summing up then, if we rigorously insist that *only* the method of partial definitions be used, nothing in the formalism obliges us to say that we are influencing the needle of an instrument when we take note of its position.

4 Discussion and return to the conventional approach

May we declare that Wheeler's approach resolves all the conceptual difficulties having to do with quantum physics? Of course not. For example, the 'indirect' measurements that can be made through correlations of the EPR type still raise thorny questions. Wheeler himself, moreover, points out that he simply tried to indicate a possible direction in which research might go, and freely admits that major difficulties remain. Our discussion in the preceding few pages has been no more than an attempt to point out the way in which, drawing on the approach Wheeler suggests, it is possible to settle *one* particular question. But it is true to say that this question does look like a conceptual embarrassment when approached from a more conventional angle.

To the difficulties which Wheeler's approach leaves unresolved the partisan of realism must add another: that his approach deals, indeed seeks to deal, only with a 'reality' that is centred on man. In his article Wheeler draws an analogy which illustrates this well. He compares the activities of the scientific community to a rather special variation on the game of 'twenty questions', which consists in trying to guess a particular word by asking of a group of friends questions of the kind 'is it an animal?', 'is it blue?' and so on. The variation thought up by the author is 'special' in that the friends of the questioner – call him John to simplify things – have agreed beforehand *not* to choose a word, each being allowed to answer as he wishes, with the sole proviso that in doing so each must have in mind a word that fits all the answers already given. As Wheeler points out, we instinctively tend to think that an electron has a definite position (and, *a fortiori*, that it exists) whether or not we are observing it – just as John thinks that the word he is trying to guess exists in the mind of his group of friends, indeed did so before he began asking questions in an attempt to discover what it is. The essential fact of the matter is that both ideas are wrong.

The full context in which Wheeler presents the analogy shows unmistakably that he considers it to be just as valid for the position of the pointer of a measuring instrument as for the position of an electron. Now, in the analogy, if John clearly represents the scientist, we may ask what is represented by the group of friends. Wheeler's answer is that 'they play the part of nature in the standard experiments'. But obviously, when he expresses himself in such a way he cannot have in mind a sense of the word 'nature' that is identifiable with a reality *knowable by physics*, since if there is one thing that is brought out into the open by the analogy it is that John does not acquire, through his questioning, knowledge of a reality pre-existing in the minds of his friends. If in this context the word 'nature' truly means anything, it can only be an external reality that is *unknowable*, or at least unknowable *as such*.

The fact remains that the idea of a nature that is fundamentally inaccessible to physics seems *a priori* rather disturbing, at least to the majority of professional physicists. Together with the remaining difficulties alluded to, this fact is sufficient to justify examining approaches that differ from Wheeler's, notably those that are more congenial to physical realism. In this section, we shall consider the conventional approach. From the point of view relevant here, we shall find it deficient in many ways – but this is a stage we must go through in our present investigative enterprise, since this deficiency will lead us to investigate, in the next section, how we can draw nearer to the way physical realism views things, taking our inspiration from the considerations elaborated by J.S. Bell in the article cited above.

Let us begin by disposing of some preliminary objections to physical realism raised by some philosophers. These philosophers posit it as self-evident that ideas can be related only to other ideas. To such an *a priori* objection, partisans of physical realism can easily answer that admittedly our descriptions must be related to experience – that is to say, to the sensations and impressions which arise in our minds – but that this in no way refutes their thesis. It is sufficient for them to point out that, as we have seen, an explicit method exists, that of counterfactual definitions, which enables us to define the properties at least of macroscopic objects (and their existence too) in such a way that, although the definitions involve experience, nevertheless the reality described by a set of such definitions can without contradiction be seen as independent of the human mind.

But disposing of this objection does not mean the way forward is clear of obstacles. To see what these are, we should first recall that if physical realism is accepted for matters having to do with the macroscopic domain, it is difficult to avoid being obliged to accept it for the microscopic domain too. Now if we *do* accept it and try to proceed, *a priori* there are two courses open to us. The first is to try to manage without introducing supplementary variables such as would provide a description of physical systems that would be more sophisticated than that provided by wave functions or kets (called 'hidden' variables according to the current convention); the second is not to exclude such variables.

The first course, again the one taken by the great majority of physicists, we shall deal with now. If we look at it closely, which has the merit of bringing out the difference of principle between it and the 'Copenhagen' approach as understood above, it is easy to see that this difference consists in using, in accordance with the general schema of physical realism, the method of counterfactual definitions. More precisely, it consists in extending this method to the definition of quantum systems and in attempting to resolve the familiar conceptual difficulties which then arise (Young's slit experiment and the like) by restricting appropriately the limits of the spheres of accessibility. At the 'technical' level this can be done quite explicitly, as we have seen, by specifying that the sphere of accessibility S corresponding to a (given value of) a property of the system *does not include* – in particular – the possible situations ('worlds') in which an instrument designed to measure a quantity not compatible with the property in question is associated with the system. With this specification the meaning of – for example – the statement that 'the momentum of the particle is p' then is that if we consider all the 'possible situations' (which of course are in general counterfactual) symbolically represented by the various points in S, and if in that set we distinguish those situations in which a device for measuring momentum is associated with the system, then in all of them the value recorded by the device is p. Once again, this analysis – which may seem a dash ponderous – merely spells out in full the meaning of the common-sense view that 'the particle has momentum p since if it were measured the reading p would be obtained'. It indicates the intellectual processes that make it seem legitimate to say that particles leaving an accelerator have momentum p, and this even when there is no device measuring that quantity actually in operation, a case in which the alternative approach would not permit such a statement to be made. In practice, such a manner of seeing things leads very directly to a conception according to which quantum states (described by wave functions or kets) are 'physically real' as understood by physical realism.

The advantages of this approach (which we could call the 'conventional' approach, for it is the one implicitly adopted by the majority of physicists) lie in its formal simplicity and efficiency. Its disadvantages, however, are serious. In the first place, it does not *truly* meet the conditions of physical realism. Rather, as we saw in Chapter 6, it is centred on weak objectivity since some of the axioms from which it proceeds do not seem capable of being expressed in terms of strong objectivity. It is thus something of a hybrid. But even if that difficulty could be avoided one way or another, there would still be others. Once wave functions or state vectors are deemed 'physically real', Schrödinger's 'cat paradox', previously avoided by the method of partial definitions, emerges again. In a sense, we could say (see the Addendum) that the entire contents of the 'quantum theory (or theories) of measurement' is simply an attempt to resolve that paradox. Here, once again, it seems that the cost of such success is the same: at one point or another (preferably where it is not obvious) reference has to be made to practical impossibility, in human terms, of carrying out this or that particularly difficult measurement. And this is why, in all rigour, it is not permissible to say that such theories reconcile physical realism with quantum physics.

As we already noted, another difficulty having to do with this approach has been pointed out by Aharonov and Albert [28]. They show that in quantum theory necessarily there are non-local physical quantities which in principle are measurable, at least in the restricted sense that it is possible to verify whether or not the physical system under investigation is in an eigenstate[5] of the physical quantity being considered. Moreover, such measuring operations leave the system in that state, so in principle the operations can be repeated as rapidly as desired. The difficulty in all this lies in the fact that then it is possible continually to check in *any* reference frame whether or not the system is still in its initial state. Under such circumstances, any ordinary measurement of a local physical quantity pertaining to the system cannot reduce[6] the quantum state other than along the hyperplane defined by the time the measurement is made, as measured in the chosen frame of reference; any other hypothesis would lead to conflict between two sets of observational predictions, both verifiable in the frame of reference in question. In consequence, the phenomenon of the reduction of the quantum state when a measurement is made cannot be seen to obey the general laws of relativity theory.[7] Again, this in no way implies a violation of relativity theory if (contrary to Einstein's preferences) this theory is seen as purely operational in character. As several writers, including the ones cited here, have demonstrated, Einsteinian causality *in the operationalist sense* is not violated here. On the other hand, the circumstances just described must be seen as implying a violation of that causality if we conceive the latter in terms of physical realism and if, in the precise sense in which the expression is used in the method of counterfactuals, we deem the quantum state to be *physically real*, and to be 'reduced' (in general) by an 'ordinary' measurement such as the one referred to above. (The theory of the relativity of states needs to be looked at separately, since no reduction is supposed to occur. In Chapter 6 we have already noted the special difficulties associated with that theory.)

The need to consider a reduction of the quantum state when a measurement is carried out has always been seen as a rather unsatisfactory aspect of the conception discussed here, according to which wave functions (or 'state vectors' or 'kets') of microsystems are deemed to be elements of reality. But what the foregoing considerations show is that this need in fact implies the existence of a genuine contradiction between that conception and relativity, itself conceived along the lines of physical realism. The difficulty we are therefore faced with can no longer be described as merely 'a rather unsatisfactory aspect'.

It cannot be said too often that the obstacles the conventional approach comes up against are *no more than* conceptual. In no way do they reduce its efficiency at the strictly scientific level. Nevertheless, it is true that they may prompt those who want to go beyond mere 'recipes' not to neglect the second of the two paths referred to earlier. This involves, as will be recalled, rejecting the *a priori* hypothesis that there are no such things as supplementary variables. It is the path we shall now explore.[8]

5 The 'BBB' point of view

These three letters are the initials of three physicists, Louis de Broglie, David Bohm and John S. Bell. With others (who never, however, amounted to more than a very small minority of the theoretical physicists), these three have long and courageously promoted the cause of a robust and uncompromising physical realism whose basic ideas would answer to the aspirations of Einstein. Like him, they take the view that such a realism must wear the colours of a mathematical realism (as defined in Chapter 1).

Like him too, they accuse the normal formulation of physics, as presently developed, of being more intersubjective than objective in their sense of the term (a sense identical with my 'strong objectivity', as defined in note 5 Chapter 1). Unlike Einstein however, they have in the field under study here devoted themselves to the task of working out in detail a theory genuinely capable of rivalling traditional quantum theory – and have met with some success.

To be able to offer a real alternative to the usual quantum theory – or to be more exact, to the usual ways of interpreting the calculation rules of that theory – means being able to generate the same predictions about the ensemble of experimental facts in the domain covered by the existing theory (as yet, the predictions generated by the existing theory have never been falsified experimentally). This is a very restrictive condition, for there is a large number of such facts. Consequently there can be no question of separate verification of all of them. Clearly, the new theory – which must be a general one, not merely a collection of models with no connection between them – must give rise to a theorem which proves that the theory's predictions about any of the facts in question are necessarily the same as those generated by quantum theory as usually formulated. Apart from a few reservations about details which are irrelevant here, it seems possible to say that the most recent versions of the new theory – in particular that provided by Bell in the article cited above – do in fact meet the condition.

This is the positive side of the matter, and there is no doubt that it is quite remarkable. Schrödinger's 'cat paradox' does not arise, at least not in its acute form. The new theories explicitly postulate the existence of supplementary ('hidden') variables in addition to the wave function, thereby permitting two particles or systems having the same wave function nonetheless to differ from one another. It follows from this that there is no danger of introducing an internal inconsistency in these theories by supposing that the objective state of the needle of an instrument is at least approximately the same before and after it is observed.

6 Discussion

The first conclusion it is possible to draw from a comparison of the approaches of Wheeler and Bell is, it seems to me, that as far as internal consistency and conceptual definitions (that is, the absence of ambiguities) are concerned, both are a good deal better than the 'conventional' approach. Further, the examination we have just made of them shows that their superiority derives from the fact that the conventional approach makes use of ill-defined rules when specifying the cases in which counterfactual implication is to be used to define properties, whereas the approaches adopted by Wheeler and Bell are precise and clear on this point. Specifically, Wheeler says 'never' while Bell says (implicitly) 'always, except of course when the so-called property in question is actually no more than a secondary one which need not be taken into account in a formulation of the fundamental laws of physics'. In the theory proposed by the latter, many of the 'observables' of ordinary quantum mechanics are in fact reduced to this simple state of secondary qualities. However, this is not the case with the positions of particles, or more precisely, with the number of fermions at each point in space. In the last analysis, we must take it that at these points the quantities in question are implicitly defined, just as they are in classical physics, by the method of counterfactual implication (for they are certainly not defined by the method of partial definitions). This is why we can say that Bell's approach accords with physical realism, whereas Wheeler's clearly does not (and whereas the conventional approach makes feeble and unsuccessful efforts to convince us that it does).

Can we now choose between these theories, and if so, which one is to be preferred? If we were strict instrumentalists we should have to choose Wheeler's; but if we went so far as to link our instrumentalism to the positivism of the Vienna School we should nevertheless be obliged, once we had chosen it, to refrain in our commentaries on it from any mention of what Wheeler calls 'nature'. This is because, as we have seen, the entity he designates by that word cannot be defined operationally and is, in truth, unknowable. Positivism as preached by the Vienna School denies any meaning to entities of that kind and so prohibits any reference to them. Since in his article Wheeler refers to 'nature' only incidentally, and given the insistence with which he rejects the idea of any 'hardware of external reality', it would seem that if we refrain from such reference we shall not be moving too far from his position. If, however, for the reasons set out in this book we decide we cannot commit ourselves to that kind of positivism – in other words, if we are determined to give to the verb 'to exist' and the noun 'reality' meanings that do not render them secondary concepts in respect of the anthropocentric concept of meaning – then once again the 'unknowability' that Wheeler's theory attributes to 'nature' – that is to say, to the very reality we are talking about – is something which may well distance us from that theory. For that reason, we may at first glance be tempted to choose the BBB approach, or more particularly, Bell's theory, which represents the most recent and to my mind most satisfactory version of it.

But, once again, along with their intellectually attractive aspects, such theories have their drawbacks too, as specialists well know. One is their lack of predictive power. They do not suggest new experiments which could put them to the test. Another more serious drawback has to do with relativistic covariance; in fact it corresponds to the difficulty in the conventional approach pointed out by Aharonov and Albert. As Bell himself emphasises, although his theory accords with the observable predictions of the theory of relativity (conceived as an operationalist theory), 'it relies heavily on the distinct separation of space-time into space and time'. In a sense, this amounts to saying that for a proponent of physical realism, relativistic covariance reduces to something of the order of mere appearance. By the same token, some instantaneous actions at a distance, non-observable but nevertheless real, can thus 'appear'.

On this last point, one very important fact of the matter is that such effects of 'non-separability' are a weakness by no means peculiar to that particular theory – which holds out the hope that future 'realist' theories may be able to eliminate them. As we have seen, recent theoretical and experimental developments (largely initiated by the same John Bell) have established that any 'realistic' theory capable of generating the same fundamental – and fully confirmed – predictions as quantum mechanics and in which reality is conceived as being in space-time (so that it describes space-time as being independent of ourselves) *must* necessarily exhibit these same particular characteristics of non-separability. Both the separation of space-time into space and time, and the non-separability inherent in the theory are nevertheless features of the latter that its author would have preferred to avoid if at all possible, since they run quite counter to the drift of the general ideas that gave rise to his theory. The foremost of these ideas (in itself irreproachable) is that a theory – or a 'vision of the world' – becomes more plausible as the 'distance' diminishes between reality as such and phenomena as they are perceived. In this respect, the fact that relativistic covariance, made manifest by so many phenomena of all kinds, can have the status only of appearance is therefore embarrassing.

Similarly, there are good grounds for uneasiness at the form non-separability takes in this theory. In theories of the BBB type, it is – first of all – the 'positions of things' that are taken as real; and given the general idea just stated, we may see there a strong *a priori* reason justifying such theories, for in fact what happens is that we always observe things as occupying defined positions (locality). Consequently, a theory will put least possible distance between reality and appearances if at every instant in time it enables us to interpret the position of everything observed or observable as being a reality in itself. However – and this is where the true difficulty lies – in addition to the 'real' things represented by the positions of particles (or more precisely by the number of fermions at each point in space), this theory is obliged to consider other 'things', no less 'real', which are deemed not directly observable, can affect the positions of particles, but are themselves highly non-local (and not expressible in terms of local quantities). What this means is that, unlike the 'numbers of particles' or the 'strengths of fields' of classical physics, none of the 'things', none of the physical quantities, in question can be quantitatively described, even in principle, by specifying their value at each point in space. These quantities are in effect nothing but generalisations of the wave functions of ordinary quantum physics, shown earlier to have in general – when more than one particle is involved – an irreducibly non-local (or 'non-separable') character.

This being so, it must be acknowledged that if the theory in question is true, the feature of locality that we perceive in things and whose re-instatement as an element of reality led us in the first place to congratulate the theory in question, in a sense turns out to be just an appearance. More precisely – though it comes practically to the same thing – it is a reality but only a partial one, which would not be worthy of note were it not for the fact that (for a reason which the theory neither explains nor even tries to explain) it is only *those* 'things' *that* are both real *and* local that we perceive, although there is a whole world of 'things' that are real but non-local – the invisible part of the iceberg – which we perceive only indirectly, through their local effects. This makes it rather difficult to claim that the theory under review here truly achieves its aim, at least if as suggested that aim is to reduce the distance between appearances and reality.

In the circumstances, even though the hypothesis that Bell's theory is an accurate representation of reality cannot be *refuted*, it seems quite reasonable to conjecture that there is a suggestion of some grand truth in the many-sided conflict between locality and quantum rules. If this conjecture is correct, we are mistaken in believing that the notions of space, time, space-time, events, and even the positions of things are 'faithful' descriptions of aspects of independent reality. Perhaps these notions are no more than mere tools for describing phenomena, that is to say, empirical reality, or the ensemble of communicable experience to put it in other words.

As I have already noted, I do not consider this conception to be one that can be imposed on us by purely philosophical considerations. I am far from adopting on this point the line of argument proposed to us by Kantians. *A posteriori*, however, in view of the difficulties of physical realism the present discussion has illustrated, I judge it reasonable to draw a distinction between empirical reality and independent reality, a distinction which is in some ways partly analogous to (but which also differs from) one of Kant's theses.

This needs clarification. The reality I call empirical – that is to say, the ensemble of phenomena – is according to this conception the only one of which the human mind can have genuine knowledge, in the sense that scientific research gives to that term. Kant would have said of this reality that 'it is in *space and time*, but this is so only because space and time are the *a priori* forms of any description of our experience'. Present-day physicists would tend to say by contrast that this reality 'is in *space-time*', and among them the disciples of Einstein would emphasise that, as space-time is not an *a priori* form of our sensibility, the very fact that physics obliges us to such an assertion shows that space-time is real, quite independently of us. For my part, I cannot go as far as that. Once again, the conclusions we have come to in the discussion so far lead me to see as more reasonable the idea that the notion of space-time (like those of locality, of events, and so on) owes a great deal to the structure of our minds, and that independent reality is not embedded in it.

On the other hand, the argument put forward by the disciples of Einstein remains a weighty one. Notions such as space-time, the curvature of space and the like are in no way familiar and cannot in any fashion be considered as constituting an innate form of the synthesis our mind performs on the data it receives from the senses. That these complex notions, rather than others, are the ones actually used by the scientific community must be due at least in part to some information received by the mind from outside. It is therefore natural to think that they reflect something about independent reality. But once again it seems to me that such notions do so only in a way that is so deformed as to make it impossible for the human mind, given the meagre information at its disposal, to reconstruct in any non-arbitrary way an image that is faithful to independent reality as it 'really' is. Such reality is for us veiled and will ever remain so.

It will be appreciated that the epithet 'veiled' here is intended to indicate a midpoint between the terms 'knowable' and 'unknowable'. In view of the argument we have just been examining, 'unknowable' is too strong a word. We should recall in this connection the situation of great conceptual arbitrariness that prevailed when Henri Poincaré put forward the philosophy of conventionalism, and also the fact that the discovery of the constraints due to relativistic covariance greatly reduced the arbitrariness that that philosophy had emphasised. That there remains something of what was thereby gained seems hard to deny. It seems reasonable, then, to think that Einstein was largely right to maintain that the structures of reality are not totally impenetrable to us. Yet 'knowable' would also be too strong a word, for it goes to the other extreme and suggests that we can know things exhaustively, as they truly are in themselves. With regard to independent reality, all the preceding discussion leads to the rejection of such an idea, or at least tends to show that it is scarcely plausible.

7 A final question

In terms of the conception of veiled reality, empirical reality comprises all that can be known scientifically, that is, with precision. That essentially is what present-day quantum physics tells us. But if that is so, a question arises: is there not a danger that within such an empirical reality – which after all corresponds to all our scientific experience – the difficulties inherent in the interpretation of quantum physics, such as Schrödinger's cat paradox and non-separability, might all reappear? If that were the case, and if the difficulties were then as serious as they would be if the distinction between independent and empirical reality is not drawn, then clearly the usefulness and therefore the validity of the distinction would be thrown in serious doubt.

The answer is that to a certain extent the difficulties remain, but they are *not* as severe as they would be if the distinction were not drawn. For example, in its strict formulation non-separability is nothing but the non-validity of the principle of separability. But, as we have seen, it is not possible even to *formulate* this principle when one wishes to keep strictly to the language appropriate to a description of mere *empirical* reality, since formulating this principle involves reference to objective states of systems. It is true that in a description of empirical reality centred on the method of partial definitions non-separability is replaced by a kind of indivisibility between system and instrument which is also non-local in certain cases. But this non-locality is scarcely more surprising than the restrictions to the method of partial definitions already made. Similarly, we have also seen that the central difficulty of quantum physics – how to avoid being obliged by its formal structure to hold that it is the observer, when he looks at the needle on an instrument, who fixes its position – can be resolved in principle. All that is needed for that is to abandon what we have called the conventional approach, with its reference to *both* the method of partial definitions and that of counterfactual implications. In particular, this difficulty loses much of its stature, at least as long as only local measuring operations are involved, once the approach (which Wheeler seems implicitly to have chosen) which consists in defining properties and states exclusively by the method of partial definitions is adopted. For such a method does not permit any counterfactual interpretation of the so-called initial state of the system (or rather, of the preparation process the system has undergone).

On the other hand, it is true that indirect measurement operations of the kind examined in Chapter 5 (those of EPR) raise some particular problems which do not seem to have been solved yet, and which could therefore cause some disquiet about the possibility of a 'physics of phenomena' that is free of all paradoxes.

It is at this stage, it seems to me, that the notion of independent reality (knowable or unknowable) can once again prove useful. For example, even if for the reasons given above I do not 'believe in' the description of independent reality set forth by Bell in his article (in the sense that I am unwilling to bet that it conforms to what *is*) I have to admit that as a model it contains no paradoxes and no internal inconsistency with regard to what this reality might be, given the data at our disposal. The mere fact that there *is* such a model is enough to dispel the disquiet just mentioned. Even if the model is not a faithful reflection of the truth, the fact that it was possible to construct it proves that in the very facts of the problem there are none that could rule out a solution. The existence of the model therefore shows us that to approach empirical reality by means of the system of calculation rules that ultimately constitutes quantum physics does not necessarily hide some fundamental irrationality in true reality. But it shows us this only if we consider empirical reality, even if it is *for us* the only one that is unambiguously accessible, given the limited capacities of the human mind, to be nevertheless not identifiable with ultimate reality. This conclusion supports, it seems to me, the conception according to which a distinction must be drawn with empirical reality on the one side and independent reality on the other.

11
Questions and answers

No one could deny that problems of the kind encountered in the last chapter are from many points of view quite fundamental. But (and no doubt largely for that reason) books that seek to examine these problems, as this one does, do not always find it easy to reconcile two imperatives: that of *clarity*, which calls for a linear exposition, passing from one idea to another by a deductive progression; and that of *exhaustiveness*, which requires that account be taken of the many and various arguments that to a greater or lesser extent influence the colour, impact, even the structure of these ideas. In practice, such arguments assume the form of observations, each calling for individual discussion, or indeed of possible objections, which require more than simply being dismissed out of hand. Inevitably, both lead to a great deal of to-ing and fro-ing, to breaks in the exposition and a plethora of asides – clearly leading to the risk that the reader will become confused and find the thread of the discourse hard to follow.

To deal with this purely technical (and hence minor) problem there is presumably no entirely satisfactory method of proceeding. The one I chose is to gather into a separate chapter – this one – the most essential of these observations and objections. Because of their nature and their great diversity, it seems appropriate to present them in separate paragraphs, with no pretence at linking them in some artificial unity. And surely the most direct way of expounding, examining and perhaps refuting them is in the form of a dialogue; indeed, I came to the conclusion that this is in fact the best method.[1] This is the reason for the unqualified use of the first person singular, as if it were always a case of answering an objection made to me personally – though in some instances this a fiction.

It would be pointless to look for a classifying principle in the questions, observations and objections examined below. The most that can be said is that the least pertinent (at least to my eyes) are dealt with first. Of these, the most important are grouped under the rubric 'crude misunderstandings'.

1 Crude misunderstandings

Question 1 I do not understand what the philosophical reasons are that drive you to insist on the existence of a *veiled* reality.

Answer Certainly, many aspects of this notion – once acquired – are agreeable to me. But there is no question for me of insisting *from the outset* on the existence of a *veiled* reality. What I do insist on is that there is *a* reality. In other words, it seems to me that, for the reasons discussed, to reject the very notion of independent reality is getting close to intellectual incoherence. But these reasons – which may, if you wish, be called philosophical, for they are not dictated to me by my scientific knowledge – do not in any way make me postulate that reality is *veiled*. Indeed, from that point of view they seem to me to be neutral. If physics were capable of constructing a complete description of human experience that was entirely expressible in terms of strong objectivity, the idea that reality is veiled would look rather dubious to me. But I discover (at some pains this time, and on the basis of the vast store of knowledge acquired no less painstakingly by my predecessors and contemporaries) that, at present at least, physics does *not* meet this condition. And I think I can also say that the chances of this happening in the future are getting smaller. The conjecture that it will never be met is therefore fairly convincing. *This* is what leads to the idea that reality is *veiled*, that is to say, knowable only as far as some of its structures are concerned, and further, without our being able to say exactly which ones of these structures are knowable (see the analogy of the recording and the concert, below).

Question 2 But, as a scientist, how can you defend a conception that is so vague? Is the notion of veiled reality not of a piece with the woolly ideas propounded by Nicholas of Cusa, by Kircher, and a host of other neo-Platonists of the sixteenth and seventeenth centuries, in short to those claims over which the rigour of Cartesianism triumphed so fortunately?
Answer There are things that must be expressed precisely. Science does that when it deals with phenomena. As a physicist, it goes without saying that I participate with enthusiasm in such meticulous and detailed work, which in its ultimate phase has room for neither woolliness nor ambiguity. But there are also things that cannot but remain vague. Rigour applied to them is false rigour; imposing precision on the description of such things can lead only to failure. You mention Cartesian rigour, but to what extent did Descartes himself keep to it when he located the seat of consciousness in the pineal gland, or when he propounded the theory of vortices and claimed it must be valid? Would he not have been better advised to see that in such matters caution was the wiser course, given the state of knowledge at the time? The other side of the coin is that of course it is just as deplorably unrigorous to say that something does not exist simply because it cannot be described precisely . . . .

Question 3 I do not understand why you attach so little weight to a proposition that most of your colleagues would accept, namely that notions which cannot be defined operationally (such as, specifically, that of independent reality) have no meaning.

Answer If you add a word to that proposition so that it reads not 'have no meaning' but 'have no *scientific* meaning' then I shall accept it quite willingly given the present state of our knowledge. As a particle physicist, I need this proposition so often in practice that I must see it as a golden rule of my discipline as it is in this century. The distinction between 'meaningful' and 'scientifically meaningful' is essential however, and the need for it is particularly obvious when the concept of reality is involved. For to deny that the notion of reality as such makes *any* sense is to risk falling headlong into the abyss of solipsism.

But that is not all. The proposition that the idea of (independent) reality has no scientific meaning was the axiomatic starting point for the operationalists. I cannot but view such an attitude as dogmatic. To justify my judgement, let me for a moment propose a counterfactual hypothesis. Suppose (what we know is not so) that present-day physics provided a rational description of all human experience and that this description could be, along the lines of the Laplacian ideal, explicated in purely mechanistic terms (and therefore in a language that would be objective in the strong sense). If that were the case, would it not be *eminently natural* (even given the fact that philosophical doubt would remain logically possible) to say that the description in question *is* a description of independent reality? In my view it would, and this, once again, is why I refuse to accept as an axiomatic starting point the proposition that the notion of such a reality is scientifically meaningless.[2] If I accept it, it is as an end point, or rather as a result of detailed scientific analysis of factual data.

Question 4 It seems inconceivable that a scientist like you could even for a moment consider abandoning rationalism, which is ultimately the only reliable guide for our thought we have.
Answer This question is linked to the one above and is susceptible to a similar reply. I have not abandoned rationalism. Quite to the contrary, I think that it is necessary to use this marvellous vehicle of thought to the limits of what it can genuinely attain, which quite comfortably include the entire world of phenomena. But I believe that these limits do exist, and that they do not coincide with the range over which we can legitimately exercise our capacity to ask questions. Incidentally, I would even go so far as to say that in general scientists, among whom I number myself here, are more conscious of these limits than are many others. They know that very many of the arguments which in non-scientific circles are presented as rational are not worthy of that name. One consequence of this is that whereas people who work in the humanities see or, rather often, believe they see in rationalism a rich field waiting to be exploited, one that offers the possibility of immensely fruitful and varied harvests, whether it be in religion, politics, or the individual or collective psyche, scientists for their part know that as long as they want to remain true to rationalism, they

must not – for deontological reasons, I would say – indulge in cogitations of this kind. For, again, they know that in fact such alleged 'advances' usually are but delusions. Now it is only *when the above-mentioned limits of rationalism have been reached* (in a specific field of course) that I think it proper to seek to go further – and then only by following paths that are necessarily our own, though we can help one another. Honesty, pure and simple, must in these final stages forbid us from describing as rational in the strict sense of the term such journeys of thought, which are in fact no more than 'reasoned' (in itself of course no mean achievement).

Question 5 I cannot understand why in what you have written you reject materialism, yet make so little reference to the brilliant 'mentalist' advances that several recent books have linked closely to the new physics. Are you unaware of the new physics? Or are you just timid?
Answer Throughout a life mostly devoted to the theory of particles my professional activity has consisted specifically, as I have already said, in contributing as best I can – as a member of a numerous band scattered over the planet – to the development of a field that is right in the van of this new physics. So I am less ignorant of such matters than most of my contemporaries. As you say, I am not a materialist; nor have I come across anything in the new physics to persuade me to become one. But I stubbornly refuse to be impaled on the dilemma of materialism versus mentalism. I take seriously what materialist neurologists tell us when they say we should be quite distrustful of the notion of 'mind', given that nowadays it is impossible to give the word any acceptable definition. It is true, then, that whoever uses it runs some risk of fogging his thought, and I am very much afraid that some of the authors you referred to have failed to avoid such a consequence – which is why I have strong reservations about their work.[3] But on the other hand, I also think it necessary to balance the neurologist's remark by the symmetric one that nowadays it is just as impossible to come up with an acceptable definition of 'matter', as physicists are well aware. Clearly, as far as we are concerned here, this puts the ideologies of 'materialism' and 'mentalism' on the same footing.

And yet from time to time physicists continue to use the convenient term 'matter'; to do without altogether would complicate proceedings considerably. But they do so fully aware of its limitations and opacity. In the same way, and for similar reasons, I believe we can and must go on talking about 'mind'. Both terms are needed, for we cannot simply leave a blank. Nor can we define everything. The two terms refer, albeit in my view confusedly, to aspects of the real that are both authentic and complementary, but not fully analysable.

It is just this incomplete analysability – or better, more or less non-analysability – of notions such as spontaneous consciousness, mind, and the like which renders for me the idea of some 'new gnosticism' based on them quite chimerical. To gain respectability, this would nowadays have to be grounded, like the atomistic materialism of our forefathers, in something pragmatic – referring to the argument that 'true is what works'. But whereas the model of materialistic atomism is readily quantifiable and is well suited to a synthetic description of many scientific phenomena that are not subject to dispute (provided, of course, it is not raised to some absolute status), the same does not hold for the mentalist model. So we must say something like this: at the level of empirical reality, science alone, in the proper sense of the term, is valid and worthy of belief. Anything esoteric, occult sciences and the like, must be sytematically cast aside. However, at the level of independent reality and the effort – somewhat hopeless but nonetheless essential – that human beings make to grasp such a reality, the notion of mind by contrast has in recent times won back some genuine respectability, not so much as a specific ingredient of knowledge than as one of the few accessible ideas that can bring human thought into proximity with the inaccessible reality.

Question 6 Do you yourself not 'mix modes'? Are not mixed modes reprehensible in themselves? Should you not, in other words, fear the same charge of 'confusionism' that is often levelled against Teilhard de Chardin?
Answer I have already partly answered this question, but a more detailed reply is warranted. Certainly there is something rather ridiculous about the eighteenth-century authors who set the physics of their time in alexandrines. More generally, any procedure that would involve incorporating in an argument in physics some non-technical considerations borrowed from a myth, a philosophical prejudice or an ideological conviction would nowadays be vigorously and rightly condemned by the scientific community, myself included.

Other more sophisticated mixing of modes are equally to be avoided. To believe, like Hegel, that large doses of pure logic will resolve some matters of fact that belong to the realm of observation is surely a mistake. But so too is to identify *without close examination* the set of the phenomena with reality in the absolute. A thousand snares therefore await anyone who proposes to venture a little beyond the science of the specialist and its well-attested procedures.

But on the other hand just such a venture is now needed. The science of the specialist has been charged with all kinds of sin, both actual and potential – wrongly, in my view. However, I am able to think that way, largely because I know the specialists who are creating the exact sciences have little in common with the 'pure specialists' whose terrifying image haunts the lay public. Even scientists have feelings, and these feelings cannot be reduced (contrary, let it be said in passing, to the view that was fashionable not long ago) to a collection of crude appetites for power and prestige. Also, these feelings include some desire for harmony between knowledge and life as it is lived. Because it is beneficial (which indeed it is, and for specialist and non-specialist alike), such a desire is quite legitimate. Satisfying it even partly or imperfectly necessarily entails some mixing of modes. Bearing this in mind, what emerges as blameworthy is not taking the risk of mixing modes (with prudence). On the contrary it is giving way to fear of the critical judgement of one's peers – and taking refuge for that reason in some purely technical matters in which the individual is protected by the forces of group cohesion.

Question 7 Should not the man of science seek above all else to conduct his reflections in such a manner as to avoid becoming lost in metaphysical concerns?
Answer In the great majority of cases, those who say such things themselves think (albeit unconsciously) in terms of a metaphysics that is both precise and unacceptable. For many of these people (I exclude theoretical physicists, for in this field they tend to be better informed), this metaphysics is mechanistic and based on near realism and the identification of weak with strong objectivity. It thus constitutes a variety of monism that cannot be reconciled with the knowledge we have today. Many who see themselves as scientific purists reveal themselves in their non-professional writings or talks on television to be metaphysicians of this crude kind (though the word 'crude' should not be misunderstood here: again, if their metaphysics accorded with physics I should be the last to reproach them for their simplicity; but their metaphysics is crude in the way the theory of phlogiston is crude).

In fact there is a method which seems to make it possible to avoid metaphysics, or at least to reduce what remains of it to a minimum. This is the method of instrumentalism, or the principle of operationalism, which as we have seen consists in taking seriously only the successful recipes, and as far as the rest is concerned, either denying that it makes any sense or resorting to the celebrated 'what do I know?' dear to Pyrrho, holding back from all answers, including negative ones. Much as I admire and sympathise with Montaigne, I confess that I find it difficult to take him literally when he invites me to follow Pyrrho in this (besides, what of Montaigne himself? Would he have been a thorough-going instrumentalist? His curiosity, his colourful way of presenting things, prompt some scepticism as regards such a conjecture . . .) Moreover, as I have said I am convinced that at this instant I exist, and this certitude, as I have also said, contributes much to the firm foundation of my conviction that there is a reality outside the domain of the purely operational. If the simple fact of having that certitude constitutes a sin of metaphysical deviationism, it is a good bet that the same sin is committed – at least under the rubric of 'sins of inattention' – by quite a number of those who accuse me of the deviationism in question . . . .

In fact, it seems to me that it is infinitely safer to acknowledge that metaphysical questions are relevant. This is not a sharp commitment since many metaphysics can be conceived of, that differ strongly from one another. When Popper for example speaks of his own metaphysics, going so far as to draw up programmes of research to advance the subject, it is clear that for all that he is no adherent of the metaphysics of St Thomas Aquinas. Although there are many issues on which I do not share the views of Sir Karl, as we have seen, in this instance I think his position is a good deal better thought out than that of people who, while sometimes appealing to his authority, yet condemn out of hand all metaphysics *en bloc*.

2 Questions with largely philosophical overtones

Question 8 Is it not one of the things science teaches us that we should seek to describe more than to explain?
Answer This is true in part, but again ambiguities should be avoided. Generally speaking, the two notions of description and of explanation are complementary. A description of raw facts is not science, any more than it is knowledge. Science begins when we seek out and discover what 'lies beneath' the raw facts. It is thus constituted by the description of these underlying things, and this description serves at the same time to explain the raw facts with which we began. In other words, it is the need for *explanation* that this *description* serves to satisfy; no progress would be made in the scientific description if there were no underlying desire for, indeed we can say need for, explanation.[4]

But here we soon run into what seems like a contradiction. On the one hand, we must acknowledge that the facts we establish are essentially 'phenomena', that is to say data about which there is intersubjective agreement. As most philosophers never tire of reminding us, it is appropriate to seek an explanation of things only among other things of the same kind (a criterion we referred to earlier) and hence to expect that the explanation of the raw observed phenomena (a spot on a screen, a bulb lighting up, and so on) will be of the same order as the phenomena themselves, the difference being that it will not be composed of *raw observed* phenomena. What then can such an explanation be? Reference – indispensable here if we are to avoid any abrupt 'mixing of modes' – to intersubjective agreement shows that it can scarcely be anything else than what present-day physicists call a 'formalism', that is to say a set of laws of an essentially operationalist character, or in other words, a set of very general rules which predict what anyone will observe in this or that set of circumstances. On the other hand however – and this is where the apparent contradiction lies – we must also acknowledge that an explanation consisting of operational rules generally leaves the mind unsatisfied. A child who wants to know what makes a car go will – quite rightly – not be content with the information on how to drive it, provided in the owner's handbook. Have physicists retained this aspect of the mind of a child more than philosophers have? That could be, and it is far from certain that it should be held against them. In other words, many physicists are in their heart of hearts not satisfied with a physical theory which reduces to just such a pure formalism. They tend to ask of a theory something quite different: they want an explanation that can be interpreted in terms of 'what really happens.'

In my judgement, their request is *a priori* legitimate. I think that merely to pronounce the maxim that we should reduce our appetite for explanation to being content with description pure and simple amounts to having, as yet, said nothing. What still remains to do is to distinguish between 'descriptions of reality' and descriptions of operational rules. The desire of those physicists who, perhaps ambitiously, aspire to the former and not just the latter makes sense to me. Be that as it may, however – and here I repeat myself in the interest of greater clarity – I conclude that given the current state of physics such a desire has a good chance of remaining unfulfilled for a long time, if not forever (though for all that I am not, and we have seen why not, rejecting the idea of reality).

Question 9 Where ultimately do you differ from Kant?
Answer This is such a difficult question because all who read Kant (to say nothing of those acquainted with this theories only by report) do not understand him in quite the same way. So all I can do here is to state briefly where my conceptions resemble or differ from *the personal idea* I have formed of Kantianism.

It is certain that as far as their *finishing points* are concerned, my conceptions and those of Kant have important similarities. For example, like Kant (and unlike Einstein it seems) I tend to see in causality (in the narrow, technical sense of the term) a structural feature of the human understanding, in other terms, an *a priori* form into which the mind finds itself constrained to insert all its experience – rather than a structural feature of independent reality. In this, I am doing no more that make explicit an attitude common to most present-day theoretical physicists and based on important features of their general theory. Another similarity is that when I examine the content of contemporary physics I find plausible Kant's idea that also space and time are no more than *a priori* forms of the human understanding, which would mean that independent reality is neither inserted into nor composed of space-time.

However, together with these similarities there are also considerable differences. The first, and the most important is that, as the two examples we have just looked at show, I seek to avoid what looks to many people nowadays a kind of dogmatism. The great ideas which Kant posits as principles I find myself rediscovering, at least in part. But precisely that: I do not posit them but rather I rediscover them, and that only after a journey of some length during which major importance is conferred on the data provided by the science of our day and to the conceptual limitations that seem clearly to follow from them. Stemming from this, it seems to me, is greater strength of conviction: for to a conception posited *a priori* by one philosopher another philosopher can always oppose (as the materialists have not failed to do against Kant) another conception, equally *a priori*. But it is harder to do that when the conception at issue is supported by specific objective facts.

A second difference, one that follows necessarily from the first, is that with regard to Kantian *a prioris* (to which I do not grant the status of absolutes, as we shall see later) I am less assertive (or peremptory, as his critics might say) than the philosopher of Königsberg. To this I feel forced – as a professional duty I might say – as a consequence of knowing of the existence of some physical models of reality (as such) which describe it as extended in space and time and which incorporate a structural causality. If one day one of these models were transformed into a theory generally accepted by the community of physicists I confess I would then have to revise my ideas rather radically. But at the moment that hypothesis does not seem likely, since for reasons given elsewhere only a tiny minority of theoretical physicists truly see any of these models as anything more than ingenious mental exercises.

A third – and very important – point where I differ from Kantianism as it is often presented is that for me independent reality is merely veiled. I do not hold that it is totally inaccessible in each and every one of its aspects, so it would be misleading to identify it purely and simply with the Kantian thing in itself, which is supposed to be completely unknowable, at least if we listen to the commentators on Kant, and Kant himself in various places. Any physicist feels keenly (I was about to say 'in his bones') the resistance offered him by what he calls, perhaps clumsily, the physical world. Infinitely more often than any philosopher, Kant included, he has occasion to note that even what seem to be the most elegant and most coherent ideas may meet with a denial that does not in the least proceed from human beings. Rather, it has to do with what he calls 'the things': with the results of an experiment he has performed or suggested to verify the ideas in question, which contrary to expectations yield data incompatible with them. Moreover, in such circumstances he is convinced, despite his lost expectations, that he has nevertheless learnt something. For him and for his peers, this kind of resistance is revealing. Revealing of what exactly? He now knows that he must guard against being too peremptory about the matter. But in view of such a 'revealing resistance' he also knows that it is impossible for him to subscribe fully to a conception according to which the independently real, the 'non-human', is unreservedly said to be so rigorously inaccessible to human beings that the very notion of it can ultimately be of very little interest to them.[5]

A fourth difference, one attributable to scientific knowledge we now have but to which Kant had no access, lies in the fact that I of course consider concepts corresponding to the *a priori* forms of our sensibility and our understanding as being applicable (for effective description) only within some domains of validity. This is so for space and time, also for the notion of a localised object, and to some extent for the concept of causality (I have in mind here the conceptual problems having to do with non-separability). In a sense, it is truly quite remarkable that the human mind should be capable of devising – albeit, as is the case, merely in the guise of weakly objective formalisms – coherent and scientifically valid models that go beyond the domains of validity of these *a priori* forms.

A fifth and final difference, this time nothing to do with physics, is that while like Kant I see that it is possible to reserve a place for what he calls 'faith' (a 'fact about our understanding which may be based on objective principles but which also requires subjective causes in the mind of the person making the judgement'), I am less exclusively concerned than Kant with morality ('practical reason') in that connection, and more so with the facts having to do with what I earlier called the 'right hemisphere' (for I did not want to use the word 'sensitivity', which is too likely to evoke the spectre of 'mixed modes', in the minds of the most critical readers, and besides is ambiguous, being used in other, radically different senses, such as the 'sensitivity' of photographic emulsions).

So we can see that if my analysis is correct modern physics differs from Kant's thought on many essential points. But even if modern physics had, somewhat miraculously, rediscovered all of Kant's principles, it would still be necessary to beware of going on to identify that occurrence with the (erroneous) claim that modern physics has nothing new to add. Once again, there is a world of difference between putting forward a thesis or idea and justifying one from the start by arguments based firmly on facts.

Question 10 That is all very well, but do not these references to the 'right hemisphere' and to 'sensibility' you just recalled constitute a backward step that is perhaps equally dangerous? Is it not true that from time to time guides or prophets arise who play upon such human faculties with a view to leading people on supposedly exalted ventures which in the event, as we have seen to our cost, turn rather nasty? And, nearer the matter in hand, should we not be disquieted by any resurgence of what Kant himself called 'lazy reasoning'?

Answer Are not the capacity to be affected and to express desires, the springs of all action, including that which consists in exercising our reason? And if the right hemisphere (you appreciate without my needing to press the point that I use that expression in a metaphorical and not a scientific sense) can, as I believe, provide each of us with intuitions, however vague, that the left hemisphere does not provide us with at all,[6] would it not be both unrealistic and – not to mince words – somewhat silly entirely to reject, in the name of reason, all that it claims to enable us to discern? Surely it would be, given that human beings are as they are; the great majority of them are not members of the Rationalist Society, nor do they envy in the slightest those who are. Nor do I suggest (pressing the matter to the limit) would it be terribly 'intelligent'; for, to take a particular example, the music of J.S. Bach is more than 'a collection of sounds agreeable to the ear'. Like many works of art or of nature, it truly makes us better informed, albeit that what it has to say cannot be fully expressed in rational terms and any attempt to do so inevitably results in mutilation and dessication.

Nonetheless, there is a good deal of truth in what you say. As I noted earlier, on matters requiring some logical unfolding of ideas it is alas extremely rare that references to notions linked with affectivity and desire (in short, with the spirit) should be made with sufficient circumspection to avoid the misunderstandings, semantic slides and indeed the crippling naivety against which you rightly put us on our guard. But this is a large and complex matter too important to summarise in a few paragraphs, and if we were to look at it in detail we would digress too far from the general theme of this book.

Question 11 Let us return to purely 'rational' problems then. If independent reality is truly as veiled as you claim, and if science is concerned solely with phenomena, I do not understand how you can say anything about this independent reality. Now is that not precisely what you are doing when you say on the one hand that non-separability has been demonstrated scientifically, and on the other that it has to do with independent reality and not with phenomena alone?

Answer By definition non-separability is, as we have seen, the violation of a 'principle of separability' (or of 'locality'), of which several slightly different formulations have been proposed. If we examine them carefully from a philosophical point of view, bearing in mind what is implicitly assumed in the terms they use, we discover that at bottom all involve a mixture of two kinds of reference. For the first part, each makes implicit reference to human experience, just as do scientific statements in general (including those that are strongly objective): for the terms they use, even if they are all supposed to correspond to reality in itself (as is the case in some of them), must needs have some relation, however indirect, with some kind of observational data. Otherwise they would obviously be quite unsupported, with as little meaning as the statement that all whatsits are whatnots. But for the second part, none of the formulations in question is *purely* operational in kind. Each (I speak of those which are valid) include one or more references to an 'objective state of the system at issue' which cannot be defined purely in operational terms (unlike the 'quantum states' of textbook physics, which are themselves defined as elements of 'recipes', in the sense Valéry gave to this term[7]).

In the formulation I gave of the principle of separability (Chapter 5) this mixing of references is explicit, since it involves both the 'results of measurements' on the one hand and on the other the parameters specifying the objective state of systems. The violation of this principle is a purely experimental fact, which shows us that a certain relationship (the one supposed by the principle) between independent reality and the human sense data is not one that truly exists. The statement 'independent reality is non-separable' is merely a simplified way of saying just that. When scrutinised with all due philosophical rigour, it can be seen as a slight misuse of language – but of the controlled kind to which the sciences, including mathematics, are accustomed and which are justified, for the specialists concerned, by the need to avoid over-long sentences. In 'vulgar' realism, which identifies independent reality with the totality of phenomena (called 'transcendental realism' by Kant) the statement in question would not be an abuse of language but rather the correct expression of the truth.

By contrast, a thorough-going positivist, or a disciple of Kant who pressed his mentor's view to extremes (and rejected in consequence some of the nuances I have suggested) would not be able to formulate the principle of non-separability in a sensible way, so that for him the thesis of non-separability would reduce to a pure and simple claim that the CHSH inequalities are violated.

Question 12 In several of your writings you have made it known that you agree with Karl Popper's rejection of the epistemological theses which he calls 'sociological'. But on the other hand you tell us that present-day physics as commonly practised in the very many specialised research institutes of the industrialised countries (excluding experimentation to do with non-separability, which is a special case) is essentially a physics of phenomena, based on weakly-objective statements which you describe to us as recipes that happen to be valid for us. Are we not thereby dropped in the middle of sociology?

Answer No, at least not to my eyes. What to me seems wrong and reprehensible about the sociological approach is not – it goes without saying – the (clearly sound) idea that the germination of theories, be they good or not, has a great deal to do with the socio-cultural soil from which they spring. Nor is it the thesis that scientific truths are essentially those about which intersubjective agreement is reached. Rather, it is an idea arrived at all too often from a pernicious extrapolation from one or other of those ideas, namely that within the science of phenomena (here I am not thinking of questions of interpretation, which have to do with philosophy) *several mutually incompatible* intersubjective agreements, relative to the same single domain of investigation are enduringly possible (even after having been submitted to objective tests) and are therefore equally valid in principle. Nothing in physics supports such a view, it seems to me, and sociological theories of knowledge, which see the final choice of a theory as depending on the results of basically subjective clashes, thus strike me as a dangerous example of confused thinking and epistemologically a truly retrograde step.

Question 13 What is the attitude of a non-materialist scientist like yourself to the claim by some philosophers that we need to develop an autonomous science of the mind ('autonomous' meaning independent of the sciences of 'matter', here physics, biology, neurology, and the disciplines associated with them)?
Answer Logic and mathematics are sciences that make no reference to the material framework which supports the human mind; they are also sciences whose structure corresponds not just contingently to the structure of our minds. In a sense therefore, the autonomy to which you allude is something quite real. But these sciences already exist. The idea of capping them, and the relevant empirical sciences, by a new discipline which would be a creation of a working party of philosophers strikes me as fanciful. Apparently the philosophers you mention are moved by the desire – praiseworthy in itself – of giving truth unshakable foundations. In substance, what they are saying is that if we assume that the mind is the child of matter, and *a fortiori* is the result of the natural evolution with its inevitably contingent and accidental features, then the creations of mind, that is to say of reason itself, will also be stamped with the same marks of the contingent and the accidental. But – they say – we should not resign ourselves to such pessimism. Rather, we should subscribe to a different hypothesis, one that should lead us to a nobler conception of what constitutes the truths of logic and mathematics, and should at the same time justify the projected autonomous science of the mind.

The idea is fine, grand even, but to me it smacks of overweening pride. The sciences are built up from below, stone by stone. They attain some height only if the craftsmen who work on them can make meticulous measurements at every stage and apply their tests precisely, to ensure that mistakes are properly corrected. In the project in question, I do not see how these conditions can be met. To the extent that we accept the line of argument advanced by the philosophers who advocate the project, it is in my opinion a further reason for conjecturing that reality is, as I think it is, indeed veiled.

Question 14 Do you take seriously the old adage that 'the whole is more than the sum of the parts'? When reading your works one sometimes gets the impression that you reduce all the sciences to physics, and even the various branches of physics to quantum mechanics alone.

Answer Yes I do take it seriously, but only after drawing all the distinctions in the absence of which its formulation is woolly and ambiguous. For the conception – out-dated in my view – that I have called physical realism, according to which science (or the sciences) describe the real as it is *in itself*, the adage can only mean that certain properties of the parts, or certain relationships between them (attraction, potential energy, and the like) which are of little importance or indeed remain merely potential when the parts are separated become essential once these parts are put together. Its scope is therefore limited. In particular, it is hard to see how the adage alone would make it possible to retain the conception of physical realism in respect of the science of macroscopic objects without *ipso facto* entailing its validity in respect of the *parts* of macroscopic objects, that is to say finally in respect of the science of the microscopic. So if we postulate that the statements constituting the sciences of the macroscopic are strongly objective, it seems a delusion to pretend that, thanks merely to the adage in question, we remove the difficulties then raised by the merely weak objectivity of normal quantum statements.

If on the other hand we agree to abandon physical realism ('transcendental realism' in Kant's language) explicitly and quite generally, then as we have seen the notion of a 'domain of validity' of concepts and theories assumes its full force. It thus becomes possible to give the adage a much stronger meaning, for every empirical science can then use its own set of concepts without, in most cases, introducing any logical inconsistency. Given that this renunciation adequately reflects my own point of view, one sees that I can subscribe to the adage without difficulty, and that in the last analysis I am not a reductionist. But it is also clear that for anyone wishing physical realism to be confirmed such a way of reasoning amounts to throwing out the baby with the bathwater.

I should add that even with physical realism discarded, merely to introduce the notion – still very general – of a domain of validity is not, as well one might think, enough in itself to guarantee that all problems of coherence that might arise will thereafter disappear automatically. The question of knowing whether the general axioms of quantum mechanics are universally valid or not cannot be answered as easily as that. It requires study of the details. Nevertheless, to people (rare among physicists but numerous in other circles) who support without nuances the thesis of separate sciences (the thesis according to which the system of laws appropriate to each domain of study are equally fundamental and that it is 'reductionist' to want to unify them), I would point out that the physical sciences at any rate did not develop in that way. On the contrary, the great progress in that domain has always consisted of or been accompanied by the coming together of diverse subjects (electricity, magnetism, optics, chemistry, for example) to form a single system of fundamental laws, precisely the one that nowadays bears the name of 'physics'. But be careful. Even if many scientists judge (boldly but not absurdly) that it is reasonable to extrapolate this first to biology and then to psychology, this does not mean they want necessarily to say that 'everything must be explained purely in terms of particle physics alone'; *that* would be equivalent to old-fashioned philosophical reductionism. The old reductionism can be pictured as a rope ladder whose bottom rung is the fundamental science of particles and fields, and the top rung the science of the phenomena of consciousness, conceived as simple 'epiphenomena'. Once again such a picture is nowadays out of date, for as we have seen the (weakly objective) science of particles is itself based on assertions which ultimately are meaningful only if referred to the group awareness of actions and observations made by the collectivity of human beings. A more accurate picturc is obtained if we make the rope ladder circular – and philosophically that changes everything.

Finally I must point out that several of the sciences we tend to think of as distinct (in the sense described above) from physics, and certainly from microphysics, readily engender in those who specialise in them a vision of the world that is very mechanistic and even 'multitudinistic' (see Question 19 below). It is a vision which, *in fine*, leads the specialist in any one such discipline to maintain that 'obviously' the beings, the organs, and so on which he studies are 'made of particles', localised and existing independently as such. These particles, he grants, are connected by forces, the details of which physicists certainly know better than he does; but 'by and large' he feels he can assume that small iron filings attracted by a magnet, or little pithballs attracting or repelling one another provide him with a model that is qualitatively correct. However, such a vision is contradicted by a non-separability which – and this is the point – is independent of quantum physics. Appealing to the hypothesis of distinct sciences is thus not enough to restore the validity of the vision just described.

Question 15 You are interested in classical philosophy and contemporary epistemology, but apparently not in current 'continental' European philosophy. This seems strange in view of the fact that to be found in this last is a wish to get beyond mere problems of language and to rediscover 'Being', a concern which your own research into reality cannot but suggest. Are you not running the risk of ignoring the existence of keys which might open to your kind of research those doors which would otherwise remain closed?
Answer The question has effectively been asked already. In particular, it has been raised by an interesting article by G. Hottois [49] on one of my earlier works. Along with his explanation of this problem, the article provides a reply which is essentially what I would give myself. In brief, the substance of it consists in noting that although initially motivated by the

hope of rediscovering Being (or more exactly, of making the idea meaningful once more), the philosophy in question and quite generally the body of contemporary hermeneutics ultimately falls back into focusing the concept of Being on questions of language. Of course Hottois presents his conclusions with rather more subtlety, but to examine what he has to say would take us too far from the theme of this book, and even further from the *methods* I am trying to use in it (but see the closing sections).

Question 16 You have spoken favourably of religions and myths, writing that they can be bearers of meaning. Is that not a position without grounds to it, and rather dangerous? In particular, is it not the case that everything which comes under the rubric of esoterism, occultism and so on is in the last analysis merely a delusion of uncritical minds, a delusion which unfortunately goes hand in hand with reprehensible commercial exploitation?

Answer The very wording of the question indicates (as I had intended) the kinship that a number of intellectuals think they see between occult on the one hand, and religion and myth on the other. When we take these latter in the way they are understood nowadays by 'enlightened' minds, such a kind of lumping together seems to me to be a mistake. We must always return to the fundamental fact that independent reality, what we call 'the real', is veiled. To the extent that scientific models constitute science, they reveal to us, though in such an indistinct and uncertain manner that it would be better to say 'they give us hints about', certain structures of the real. And given that, I cannot see on what basis we could maintain that religion and myth are not themselves also 'models', giving us – in a manner equally indistinct and uncertain – access to *other* features of the real. With regard to independent reality, the uncertainty is, I insist, just as great in both cases. On the other hand, I see very good reasons, evident to all scientists, for maintaining that only the scientific model, to the exclusion of religion and myth, can give us access to empirical reality, that is to say, to phenomena and therefore to the springs and levers of action. So I deny – once more – that there is any value in the 'occult sciences', and am adamant that anyone claiming to draw from my writings any argument which tends to establish their validity is quite mistaken.

Question 17 In your writings you are fond of referring to Spinoza. What do you think of pantheism, and are you a pantheist?

Answer If by pantheism what is meant is some dispersal of the divine in phenomena, I am not at all sure – to say the least – that Spinoza may rightly be called a pantheist, as he often is. While he indeed wrote of 'Deus *sive* Natura', he also specified that he was talking about '*Natura naturans*' rather than '*Natura naturata*'. Only the latter can be identified with the totality of phenomena.

As far as I am concerned, in any case, I decline the pantheist label. Pantheists, of whom there are many nowadays, seem to me to be guilty of a serious confusion. In my terms, they do not distinguish between the real in itself (Being) and empirical reality. That goes against what I take to be the truth.

3 Questions bearing in some way on technical aspects of physics

Question 18 You attach great importance to the distinction between strong and weak objectivity. But is not that distinction questionable? Every science, even every special discipline within a science, has its object. That of the quantum theory of measurement is the study, not of things nor properly speaking even of physical entities, but rather of the results of measurements and of the way these measurements are obtained under various circumstances. The objectivity of the theory involved cannot therefore be called 'weak' by contrast with that of theories dealing with the physical entities themselves, that would be described as 'strong'. The only difference between the quantum theory of measurement and such and such other theory is that they have different objects, or 'referents'.

Besides, as you yourself have already acknowledged, even strongly objective theories (to use your language) refer in some way to experience. Given all this, how do strongly objective theories differ from those that are only weakly objective?

Answer I have already sketched a reply to the first part of this question, but perhaps we should further explain its content. What seems to me the most important, indeed the crucial point is as follows. It is definitely *not* the case that we have on the one hand a quantum physics dealing exclusively with physical entities the reality of which would be no more problematic than that, for instance, of the electromagnetic or gravitational fields in classical physics, and on the other hand a theory of the measurement operation whose object would be limited in the same way as is that of the theory of the precision of measurements of classical quantities. What *is* the case is that it is already at the level of the very axioms of quantum physics in general that the notion of measurement (and indeed the corresponding notion of the preparation of systems) comes into play.

What is more, it is very difficult to interpret this latter fact as a mere linguistic convenience to which textbooks would resort just for conciseness for that would imply that more careful formulations would make it possible to avoid the notion in question coming into play. What happens in fact is precisely the opposite. Concerning for example, the role played by the notion of system preparation in the quantum axioms, in the most elementary formulations this role is somewhat obscured by the use made there of the word 'state'. But in the work of mathematical physicists who seek to render the axioms as fully consistent and widely applicable as possible (in particular, to render them compatible with the requirements of relativity), it is often pointed out that for clarity the notion of a quantum state should be replaced with that of preparation of systems, or at least be understood as reducing to the latter.[8] Now clearly the notion of system preparation is, like that of measurement, only weakly objective. So we can see that here we are dealing with a discipline – 'sophisticated' quantum mechanics – whose object is no different in kind from that of physics in general, indeed is identical to it, but that nevertheless, in statements of its fundamental principles, involves reference to operations carried out by the collectivity of human beings.

No doubt someone will point out that this preparation of systems and these measurements of observables can be made using automatic instruments. That is true. But these instruments are themselves made of atoms, that is to say, of quantum systems. As soon as we propose leaving completely out of consideration the human element in the preparation of these instruments, and to disregard the plan which prompted that preparation, we find ourselves conceptually obliged to take into consideration only the complex system comprising all the atoms of these instruments (and possibly of their environment) *plus* the atom (or the quantum system) they are 'preparing' and on which they are to carry out 'measurements'. Now normally any system of atoms comes under the jurisdiction of quantum mechanics. So far, despite valiant efforts, physicists have not been able to 'put their finger on' any exception to this simple rule. We thus see that, from whatever angle we approach the question, it is difficult to avoid a kind of universalisation of quantum mechanics, that is to say, of a theory that is only weakly objective . . . .[9]

Now concerning the second part of the question, to me it seems a sufficient reply to note a necessary distinction. There are statements which are *interpretable* in terms of reality in itself, and there are others, essentially in the form of rules or 'recipes', that are not. Certainly, with the exception of purely logical or mathematical statements, all scientific statements make some reference to experience. But it is nonetheless true that the first kind may be thought of as describing a reality from which man is absent, while the second then lose all meaning.

Here it seems appropriate to emphasise with some insistence that if we are to have clear and unambiguous definitions and explanations in contemporary physics the distinction between strong and weak objectivity is truly indispensable. One of the most striking examples of this (already alluded to in passing) has to do with defining the concept of causality within such a physics. If we try to define it in the language of strong objectivity, non-separability and the violation of Bell's inequalities soon convince us that such causality is subject to violations, which render the definition in question unusable, or at best ambiguous and thus dangerous to use. Now this is a worrisome conclusion, for technically the concept remains one of the most useful. Fortunately there is a way of getting around the problem, one that well-informed physicists – and also those who are guided on the matter by a kind of instinct – know well. Specifically, it consists in adopting a definition of causality framed in a statement that is only weakly objective: that is to say, in this case, one that refers explicitly to the results that can be obtained if one performs some manoeuvre or other under some condition or other. This is not the place for technical explanations but specialists are well aware that – particularly in the domain of relativistic quantum physics, which is extremely delicate – it is the only known way of avoiding paradoxes and saving all that is valuable in the idea of causality.[10]

Question 19 You place great emphasis on quantum mechanics (relativistic or non-relativistic), that is to say on a theory that has already been in existence for half a century. On the other hand, you have little to say of theories that are more recent and deal also with the atomic and sub-atomic worlds, such as the quantum theory of fields, the theory of the S matrix, gauge theory, unification theories, theories of quarks, gluons and the like. Now these theories make up the principal themes of current research in laboratories and institutes devoted to theoretical and particle physics, in particular those conducted by the group to which professionally you belong. Does that not call for some explanation? Again, you do not make use of the major discoveries in astrophysics. In addition – and this follows from that – you insist in your writings that quantum mechanics refutes what you have called 'multitudinism', that is to say a physical realism in which the principle of separability is said to be met and which purports to describe the physical world as essentially composed of myriads of 'little bits' (atoms, particles, the technical name does not matter). Now the most recent theories of the kind in question – notably that of quarks – seem, at first glance at least, rather to conform to the multitudinist conception. What are we to make of this?
Answer It is quite natural that the objections contained in the question as a whole should come to the minds of many readers, principally non-physicists. And it is important that they grasp the reasons why these objections are unfounded.

Explaining the matter to non-experts is not easy because it involves the very structure of these theories. The main point, which needs to be grasped from the start, is the distinction between a theoretical framework and a theory in the ordinary sense of the term. Most fortunately, the history of science furnishes us with a clear and straightforward example of this distinction, to do with the two most famous discoveries of Isaac Newton. One of these, known as Newtonian mechanics, is a classic example of a theoretical framework. The three Newtonian laws which constitute it are laws which on the one hand are held to be perfectly general (they are supposed to apply to all phenomena), and on the other (accordingly) do not specify the 'law of forces'. Newton's second great discovery, the law of gravity, is by contrast an example of what I just called, for want of a better term, a 'theory in the ordinary sense of the term'. The law of forces – the force is proportional to the product of the masses and inversely proportional to the square of the distance between them – is there well specified. Over the two centuries or so after its appearance, Newtonian mechanics was taken to be absolutely true, while forces other than gravitation and obeying more complex laws than that of gravitational attraction were discovered. The theory of gravitation was thus revealed as a construction which concerned *some* phenomena but not others. *All* phenomena, however, including those for which the law of forces was not the simple one devised by Newton, were still supposed to fit exactly into the framework of Newtonian mechanics. In particular, attraction was still thought of as proportional to force, whatever might be the law that governed that force (with a proportionality coefficient, the inverse of the mass, that was held to be a constant of the object). The difference is clear.

Once this distinction between a theoretical framework and a theory in the ordinary sense is clear and properly understood – such as by means of this example – the next thing to note is that this distinction has a new point of application in present-day physics. It would not be going too far to say that quantum mechanics, or more exactly the set of the general principles of quantum mechanics, constitute a theoretical framework (which we can call the 'quantum framework'), whereas modern atomic theory and also theories which came to light more recently, from the quantum theory of fields to the theory of quarks, gauge theories and so on (and *a fortiori* the great theories of astrophysics) are 'theories in the ordinary sense of the term', which fit into the quantum framework. What shows that this way of seeing things is correct is the fact that, like Newtonian mechanics in its day, the quantum framework does not in any way entail a 'law of forces' and that, in whatever way the new theories represent or rather generalise the concept of force, they all, without exception, presuppose the validity of the quantum framework (the principle of superposition and so on).[11],[12]

We can now see, I think, that a framework theory is something much more fundamental than the theories that can be fitted into it. Epistemologically, the most important problems are those, if any, that are raised by the framework theory. In the particular case of the quantum framework, a more technical study of the various recent 'theories in the ordinary sense' I just mentioned fully confirms that conclusion. This is not the place to describe such a study, but when it is carried out it becomes apparent that any attempt to get to the heart of the subject matter of these theories inevitably leads beyond the questions peculiar to each of them; we always encounter in them the same fundamental epistemological problems, which not surprisingly, are those related to the quantum framework.

I hope that what I have just said provides sufficient explanation of why a particle physicist such as myself has been led to occupy himself principally with the quantum framework when dealing with fundamental questions. It remains for me to make it apparent (without using mathematical formulae of course, though this makes it more difficult) why, given the role of theoretical framework played by the ensemble of quantum principles, multitudinism is to be ruled out. To do this, we must first note that contrary to appearances, the emergence of the theory of quarks in no way changes the basic facts of the problem. Certainly, the new notion of quarks shows us that the nucleons, which we used to think of as simple, do in fact have a structure; it also shows that the *corpuscular model* is, within a very specific domain of validity, appropriate for the qualitative description of certain important aspects of the physics of this structure. But with regard to these (extremely general) aspects however, that discovery has nothing truly specific to tell us. In this it is analogous to the discovery, made at the beginning of this century, that the atom has a structure which can be appropriately described, in qualitative terms, by the corpuscular model of electrons and nucleons. Indeed, we know that the model goes beyond the purely qualitative, since it is possible to write an equation for the behaviour of the electrons, solve it and verify quantitatively that it agrees with experiment. To some extent, unfortunately much less, the same holds for quarks. But – as already noted – in *both* cases the equation is written on the basis of quantum principles, and the very meaning of the symbols used can be defined only by reference to them. Granted that quantum mechanics refutes multitudinism we thus see just how simplistic it would be to think that the discovery of quarks could in any way allow us to claim that 'multitudinism' had merely fallen into a *temporary* disrepute through the discovery of the quantum theory of electrons.

That being said, it must be acknowledged that the corpuscular model – which *in its first approximation* involves a kind of multitudinism – works so well that even the theoretical physicist finds it hard not to resort to it. Standard expressions such as 'particle physics', 'theory of elementary particles' and the like illustrate the extent to which our intuitive picture of for example electrons, which we liken to tiny grains of some kind, is so fruitful. How indeed could we do without it, considering that it enables us to simplify our discourse to an extraordinary degree (otherwise it would be necessary after every line to come back to a description of the apparatus), and that *almost everything* occurs as if this intuitive picture were an authentic description of reality in itself? To describe a 'collision experiment' (already the word 'collision' invokes our model!) without speaking of *the* particle, of a certain type, which deviates to the left, or *the* particle, of another type, which deviates to the right, and so on would be just about impossible. In the vast majority of cases we can speak in such terms quite satisfactorily, attributing in thought well-defined intrinsic properties to each particle thus specified. We can do this because, in this majority of cases, the kinds of reification – unjustified, according to quantum mechanics – we yield to when we thus ascribe definite properties to each of the particles under consideration turns out (according to the same quantum rules) to have no harmful consequences as far as the observational predictions we can make are concerned.

But it is clear that if such modes of description turned out to be exact copies of what reality in itself actually *is*, it would not only be in the majority of cases but in *every* case without exception (meaning whatever the kind of particle involved) that *at least one* detailed description which conformed to them would have to be compatible with the predictions generated by the theoretical formalism and with the observed facts. This however is not the case. Some kinds of particle ('neutral K mesons' are the best known of them) do exist for which quantum mechanics generates predictions that are in fact grossly incompatible with the very simplistic realist picture just sketched. And in such cases it is the prediction of quantum mechanics, not that which might be derived from the picture, that experiment proves to be correct. This should help to dispel the illusion that a naive 'corpuscular' picture can be elevated to an authentic description of that which truly is.

What happens here is just what has happened in the case of many delusions. As long as the horizon of the medieval serf was limited to his own parish, he could believe the Earth was flat. Nothing in his everyday experience contradicted this conception, which had the advantage of being very simple and fully accessible to his mind. The theory that the Earth is round, however, explains all his experience equally well, given the orders of magnitude involved, *and* several others as well, which are incompatible with the first theory. That, ultimately, is why it is correct. Besides – this is a point of logic that needs stressing – it is correct in a general way. It would clearly be absurd to say that the Earth is round for Magellan but flat for the serf who has never left his fields. To a physicist, it would be about as absurd (despite the differences) to say that the naive corpuscular model holds good for all particles except the neutral K meson, for which it is false.

We might wonder, however, whether there might not be *sophisticated* ways of interpreting the quantum framework which would make it compatible with retention of the notion of particles, indeed even with the notion of a multitude of particles. The answer is a qualified yes. There is in fact a formalism for describing the quantum framework that has proved both powerful and fruitful, for which we are indebted to Richard Feynman. Feynman makes much use of the word 'particle', and even of the picture of a 'corpuscular trajectory'. At first glance this looks like a return – as far as principles are concerned – to a kind of corpuscular realism in the classical sense. A closer study however shows that this is not so. One of the essential features of the formalism in question is the distinction it draws between the case where, to calculate the total probability of a given event, it prescribes adding the *probabilities* of the various ways in which the event can occur, and the case in which, to find the same quantity, it prescribes adding certain other entities, called the *probability amplitudes*. Now this major distinction is, within the formalism, ultimately based on the *observability* or *non-observability* of what it is in which these 'various ways' differ. In other words, it is not – no more than is that which corresponds to it in ordinary quantum theory – specifiable by any statement that is objective in the strong sense; or more specifically, by any statement that does not involve the notion of 'instruments defined as such by the fact that they are used as instruments'.

It should be added that in cases where probability amplitudes and not the probabilities themselves are to be summed, we should not deceive ourselves about what 'corpuscle' and 'trajectory' mean. If we define the notion of probability in such a way that it truly corresponds to experiment, that is to say, as the 'limit of frequency', then for an event which can be brought about in a number of compossible but mutually exclusive ways, the total number of occurrences in repeated experiments is inevitably the sum of the number of times it comes about in each of the possible ways. As a result, under these conditions what is to be summed in order to obtain the total probability required is inevitably the probabilities and nothing else. To posit the rule of summation of probability *amplitudes* amounts in fact to abandoning, though without saying so, the realist content of the idea of trajectories (or more generally of paths), and to keep just the linguistic expression. Under such conditions, it is clear that corpuscular multitudinism (the only multitudinism *a priori* envisageable within this particular formalism) is in fact ruled out, since it presupposes both strong objectivity and the validity of the notions of corpuscle and trajectory.

Question 20 Must we conclude from that that no physical theory exists which is capable of reconciling the totality of phenomena with a demand for strong objectivity?

Answer Strictly speaking, to make the claim in such a hard and fast way would be to distort the truth. Once again, if one looks hard enough one can find in the specialised scientific literature some consistent theoretical schemes whose assertions do conform to the prescriptions of strong objectivity. Here I am thinking in particular of theories of what I called above the 'BBB' kind, whose major ideas were described in Chapter 10. (For the sake of completeness mention should also be made of some particular models; these limit their ambitions to providing a mechanistic description of this or that experimental fact which happens to cause problems in respect of strong objectivity. But they do not make any claim to be generalisable over the totality of microscopic phenomena, nor can one conceive how they could. For that very reason they do not constitute the makings of the kind of theory we are interested in here.) I also explained there why these theories met with the scepticism of the majority of theoretical physicists. Briefly, it is because on the one hand their considerable complexity is not compensated by any greater experimental fruitfulness, while on the other hand the known relativistic invariance of the basic physical equations does not seem to correspond in them to a fundamental property of 'that which is' but rather is merely an 'appearance' resulting from some kind of conspiring together of various effects. This marks something of a return to pre-Einsteinian ideas (one thinks for example of Lorentz envisaging at the beginning of this century that bodies in motion may experience the physical effect of contraction) that all theoreticians, including the authors of such schemes themselves, agree is unsatisfactory. In the eyes of some of the latter, however, the unsatisfactory character of their schemes is compensated by the advantage of recovering strong objectivity. But the majority of theoretical physicists see it otherwise, judging that such an advantage is too philosophical to be of much weight and so does not compensate for the original disadvantage of relativistic non-invariance.

Quesion 21 From that point of view, what is the position of the theory of the relativity of states?
Answer Yes, indeed one can say that it too is in a sense a strongly objective theory, but if the reader just has another look at the summary of it presented in Chapter 6 he will easily be convinced that it cannot be fitted into the multitudinist framework.

Question 22 In connection with the proof of non-separability we may ask ourselves two questions. First: since the quantum formalism already entails non-separability, and given the fact that this formalism has been corroborated experimentally, what need was there to look for further proof? Taking recourse to the violation of the Bell-CHSH inequalities seems both complicated and superfluous. Second: elementary quantum mechanics admittedly leads us to expect correlations incompatible with some unavoidable consequences of the principle of non-separability (specifically, the Bell-CHSH inequalities). But as we know it is not relativistic. On the other hand, the most powerful arguments for the validity of the principle of non-separability are relativistic (since they are based on the finite speed at which influences are propagated). Can we not contend, on the basis of that observation, that in reality there is no proven contradiction between quantum mechanics and the principle of separability, and that the two simply concern different domains?
Answer The first question has already been answered – in the negative – in the account on non-separability presented in Chapter 5. Briefly, what justifies that negative reply is the fact that in what concerns fundamental conceptual principles it seems arbitrary to *postulate* that quantum mech-

anics is complete; but, as we have seen, for non-separability to be seen as simply a consequence of quantum axioms, or in other words, to be able to 'short-circuit' consideration of the Bell–CHSH inequalities, it is necessary to postulate precisely that.

As for the second question, it is answered as follows. On the one hand, it is not the case that the plausibility arguments which *a priori* tend to make us see the principle of separability as valid – and indeed obvious – are exclusively relativistic. The truth is more subtle. Indeed, if we refer back to Chapter 5 we shall observe that relativistic considerations are not put at the forefront of the matter and that despite this the plausibility arguments presented there appear quite strong. (The choice of such a mode of presentations is not fortuitous; in fact there are certain difficulties in basing the principle of separability on essentially relativistic considerations; they are described fully in [2A].

On the other hand, it is not true that the quantum predictions to which the Bell–CHSH inequalities are compared are exclusively relevant to non-relativistic quantum mechanics. Earlier we drew a distinction between a 'theoretical framework' and an 'ordinary' theory, and noted that the fundamental quantum principles constitute a theoretical framework, which we called the quantum framework. Now this quantum framework is neither relativistic nor non-relativistic. Non-relativistic quantum mechanics can be fitted to it by supplementing it with the Schrödinger equation, and it can take on an essentially relativistic character by being supplemented by other principles (of which the most important bears the name 'micro-causality'; though well known to specialists, there has been no need to bring it to bear here, and hence no call to describe it in any detail). Now the quantum predictions relevant here are essentially a matter of the quantum framework alone (together, of course, with the initial specifications). They are therefore not as limited in scope as the question would lead us to suppose, which thus deals with the objection.

Question 23 Let us end with a question which mainly concerns physics but is also of more general scope. An epoch is now closing whose thought has been marked by a kind of reasoned faith both in the existence of universally valid knowledge and in the supreme value of such knowledge. Nowadays we find philosophers whose excellence is reflected in the size of their audience expounding the idea that the notion of the 'universal' is simplistic and should be banned. The arguments they develop refer in part to recent scientific advances which make it clear how complex reality is. What do you think of this?

Answer In my opinion, the developments you mention are important in two respects. The first is the emphasis they give to progress recently made in the study of phenomena that are both complex and out of equilibrium. Research in this domain, which has to do with mathematical physics, is being pursued very actively at present. Already it can be said to have made clear to us a richness in the idea of complexity which hitherto was perceived only dimly but which is now leading us to relativise, in some way, the importance attributed to the system of laws that govern the elementary. In many circumstances, these laws, though applicable in the abstract, cannot in practice be used to explain the phenomena under study. Those which *de facto* serve this function are more specific to a particular kind of problem and less universalisable in consequence.

The other interesting aspect of the developments in question is much more philosophical, perhaps even sociological, in character, but it is related to the first. It has to do with a certain discredit rightly cast on an improperly simplified conception of the universal, which not long ago was widespread. In the domain of economics, sociology or politics for example, where by their very nature the phenomena are extremely complex, this conception involved the application of untested schematisations leading to the formulation of very simple laws, which were then elevated all too quickly to universal principles. While it looked scientific, such a procedure is in fact the very opposite of the true methods of science, in which any approximations that we are led to make are tested with great care. The severe criticism now being made of the procedure can only be for the good.

Nevertheless, oversimplification must be avoided in formulating the criticisms as well. Simply to deny that fundamental universal laws exist would undoubtedly be to do just that. Taking account of Maxwell's equations is essential for anyone who wishes to understand something of the behaviour of the flux of charged particles known to exist in distant galaxies. The study of nuclear physics and the physics of elementary particles provides a good understanding of the way stars work in general, and also of the pecularities of certain classes of them. It would be quite absurd to close our eyes to such facts, which are evident proof of universality. It would therefore be naive – and by no means a 'philosophical breakthrough' – to reject the universal to such an extent that we returned to some kind of primitive paganism, according to which, say, each pulsar star would harbour a nymph matching the period of her dwelling place to the periodicity of her moods . . . . For intellectuals, always sensitive to fashions of thought, there has lately been the danger of a similar misdirection of their capacity for reflection. But in such fields, most of the time one excess corrects another. So there is no real cause for concern in these tendencies: rather we should look to their positive aspects, described and underlined above.

12
Summary and perspectives

Given the complexity of the relationships between present-day physics and epistemology, it seems useful in this concluding chapter to consider once more the salient points of the preceding analysis, not only to present them in some readily accessible form but also, more importantly, to allow us as we proceed to examine some possible extensions.

We must be quite clear at the outset that the latter undertaking necessarily involves crossing the border between what is firmly established and what is merely possible. *A priori*, reputable scientists have reservations about this. They are aware that in their profession crossing that imaginary line is somewhat rash, and they observe that their colleagues who do so are sometimes on the verge of delusion. Quite rightly, this confirms them in their caution. Paradoxically however, it is nowadays the case that refusing to make *any* move of such a kind is also dangerous, and in extreme cases can lead to intellectual incoherence.

With hindsight, this is what the history of positivism, both that of philosophers and of physicists, finally shows rather clearly. By refusing to allow that there is meaning in the notion of independent reality – on the grounds that it was impossible to construct an operational definition of it; by introducing the notion of a 'linguistic framework' – in order to mitigate the disadvantages consequent on their first move; the positivists of the Vienna Circle arrived at a philosophy which it is appropriate, using the adjective they made their own, to call a 'purely linguistic standpoint'. And though they were verbally opposed to several theses of the positivism of the philosophers, the physicists of the Copenhagen School, for their part, built up a quantum mechanics in which certain lines of reasoning when followed closely suggested, as we have seen, rather similar views. Now this purely linguistic standpoint, in which every care is taken to make sure that the border we have been talking about is never crossed, itself ran into difficulties. In earlier parts of this book we examined the more technical characteristics of these; in view of what is to follow, however, it would be as well to recall the general characteristics of two in particular.

The first is related to the development of science itself and becomes clear as soon as we compare the purely linguistic standpoint with traditional conceptions. In the latter, the world is seen as having its own existence in itself and therefore its own system of laws; consequently, an investigator subscribing to such conceptions spontaneously draws a sharp distinction between laws and methods. For him, laws have the status of givens; obviously then there is no question that he could change them in any way; whereas methods, by contrast, are free for him to invent. But for an adept convinced of the truth of the purely linguistic standpoint, such a distinction can by no means be radical, for there is no objective foundation on which it rests. Such a person has no real need to distinguish between on the one hand laws of nature and on the other the methods – approximations in general – by means of which he and his colleagues work out their observational predictions. As a consequence of these views, the algorithms generated by a generally applicable method of calculation can gradually be reified as elements of a 'description of nature' without the theoreticians responsible being themselves truly aware of the conceptual slide that is involved. In contemporary physics a striking example of this kind of development is provided by the concept of the virtual state. This concept stems from the method of approximations known under the name of 'method of perturbations'. Initially the latter was just an approximate way of obtaining, in the case of weak interactions between systems, acceptable estimates of quantities that cannot – or cannot yet – be calculated exactly. (In the simplest cases the quantities in question are solutions to known equations which cannot be solved analytically.) But it turns out that this particular method brings into consideration certain functions which in form are partly analogous to those which serve to describe the quantum states of systems. This led to the expression *virtual states* being associated with these functions; using that expression made it possible to shorten and hence clarify accounts of the application of the method. Finally, through this progressive slide, the notion of a virtual state (of quantum systems, particles and so on) took its place among those deemed by physicists as being 'fit and proper' for describing situations that exist in nature (in more or less the same way as the notion of a real state is deemed to do). At the technical level, such a development is all to the good if, as in this case, it results in notions which fit into a framework which is quite general. But the ease with which it enables us to construct 'natural entities' is obviously fraught with danger – notably the danger of a proliferation of partial models and procedures elevated to the status of things, forming such a swarm that any substantive vision, however fleeting, of the real may well be lost.

A second objection of a general kind can perhaps be raised against the purely linguistic conception. If one day this conception were by some mischance to be taken truly seriously, it would be such as to deprive scientists of some of their main motivations. When a palaeontologist, for example, looks for the bones of a diplodocus, the reason for his quest is, as has been noted, the conviction that such creatures have *really* existed. And here the word 'really' quite obviously means 'independently of the scientist and those like him'. If our palaeontologist were won over to the purely linguistic standpoint, and persuaded himself in consequence that the only valid reason for his research is that it might enable him to write scientific articles whose content harmonised with that of other scientific articles, he might well lose heart and change his job.[1] Similarly, particle physicists who in their heart of hearts have convinced themselves that the only real reason they could have for constructing large accelerators is to give themselves rare sensations whose coherence can be verified intersubjectively would have a hard time of it, trying to put a convincing case for research funds from financing committees.

Such considerations may seem rather trivial. But in what follows the conclusion derived from them will play a role that is not insignificant; hence the need to spell things out here. Be that as it may, it should now be apparent that in conjunction with the more technical and less immediately engaging arguments developed in earlier chapters, such considerations have some force. In other words, I shall now assume that the need to escape the ethereal void of the purely linguistic standpoint and to accept the notion of a reality external to man and not dependent upon him is recognised.[2]

1 The totality of experience and the totality of the real

As the study of physics shows with particular clarity, science is a theoretico-experimental construct. In a first characterisation, this means that its base is experience and that theory is the cement that binds together the many components of that experience. In a second, science may be seen, no doubt rightly, as the model which most closely approaches the ideal (attainable or not) of a body of knowledge firmly based and free from the play of opinion and mood.

The first characterisation of science makes it quite natural to identify the domain of science with that of communicable experience. The second suggests that the entire domain of *true* knowledge may well coincide with the totality of experience, indeed ultimately with the ensemble of all assertions that are verifiable and have been verified.

If from (say) the beginning of the nineteenth century none of the many changes in conceptual interpretations (what sociologists of knowledge call 'scientific revolutions') had taken place, and if, moreover, it were still possible to formulate in strongly objective language the content of the whole of science (especially that of physics, such an important part of it); then the two characterisations just mentioned could be readily and 'naturally' interpreted in the realist terms still used in popular treatises. Like Laplace before us, we would simply say that the ensemble of verifiable and verified claims anchored in experience and constituting science *is, in principle, adequate to describe the totality of the real.* Of course it would still be logically possible to retain philosophical doubt of the kind based on Descartes' hypothesis of the deceitful demon. But given the efficacy of both our science and our everyday representation of 'things', to entertain such doubt would seem more or less like intellectual gymnastics. A respectable person would 'leave such things to the philosophers'.

But as we know these suppositions are contrary to the facts. Over the last two centuries the growth in the scope of physics and the elaborateness of its formalisms, although harmonious, has been accompanied by considerable upheavals in the conceptual foundations of theory; it is obvious that this greatly affects the value a realist in quest of *true* concepts can now attach to the set of those which *currently* are proving fruitful. To this argument of a general kind – which could, it is true, be countered by claiming that only now has science come of age – can be added the fact that it is precisely within the formalism of the 'mature' physics of today that it is proving hard to see, among the 'algorithms' – particles, waves, statistical operators, S matrices, positive linear functionals and so on – used by theory to synthesise experience, just which ones it is appropriate to prefer over others to receive the epithet 'real' without thereby affecting the efficacy of the theory. It is scarcely surprising that this is closely linked to the weak objectivity which, as we have seen, is proper to the formalism in question. The feature of the theory that it is merely 'weakly objective' can admittedly be eliminated, but only at the price of having to reckon with speculations buzzing with 'dissonances', and between which it is is impossible to choose in non-arbitrary fashion. Hence in the current state of physics, unless we have recourse to the purely linguistic standpoint, we are obliged to draw a distinction, at least formally, between independent reality and the totality of experience. For the latter, the name 'empirical reality' has been chosen here.

This clearly has wide-reaching philosophical implications. Before tackling them, we should first note that apart from any specific philosophical or *a fortiori* metaphysical purpose, such a distinction is already useful to the scientist whose sole aim is to understand the opposing arguments of other scientists. It is undeniable, for example, that it illuminates very well the true nature of the differences between the approaches of Bell and of Wheeler, examined above. The objective of Bell's approach is independent reality (which he is optimistic enough to deem knowable . . .), while Wheeler's approach can perhaps be interpreted as having the more modest objective of describing empirical reality. Similarly, the debate about determinism in which Ilya Prigogine and René Thom [22] not long ago found themselves on opposite sides becomes a good deal clearer if we appeal to this distinction. To a certain extent, the two can be reconciled simply by noting that they are not talking about the same things: Prigogine is interested in empirical reality, and in that context is right to stress the role of chance and the importance of complexity; whereas Thom, like Bell, is interested in independent reality and simply cannot, at that level, be criticised for adopting the determinist thesis since all previous serious attempts at theories which purport to describe such a reality have all effectively been determinist. Similarly, the distinction in question greatly clarifies the debate which occasionally opposes some physicists who talk of deducing from the equations of their science the existence of a causality that is instantaneous at a distance, – or indeed, as with Costa de Beauregard, of a causality that is genuinely retrogressive – to other physicists who maintain that this is illusory. Here too both sides are to a large extent correct, and there is nothing paradoxical about it because their objectives are quite different. When for example the former point out that non-separability implies 'instantaneous' causation, and that it therefore follows that if the observable predictions of quantum mechanics are true then *this* causality – relations between events in themselves – must exist, they are correct. The point is indeed proved. But when the latter, who define causality at a distance as the possibility of transmitting *usable* signals, deduce from the same quantum mechanics the non-existence of instantaneous causality thus defined, they are not wrong, for that too can be proved (see for example [2A]).

It can be seen from this that in current theoretical physics many of the debates that create divisions among physicists, all of whom are convinced that they conduct their reasonings within the narrow framework of pure science, are in truth debates that go beyond this framework. A little philosophy could resolve their differences. But if that is to be achieved it is necessary that partisans of mathematical realism refrain from elevating to an absolute dogma the principle that the real is totally intelligible. And it is also necessary, by the same token, that those physicists who assign to science the more modest objective of simply describing phenomena (synthetically and mathematically, of course) refrain from systematically condemning (by implicit reference to the word 'metaphysics' applied in the spirit of the Vienna Circle, namely as a synonym for a nullity, something absurd and worthless) the idea that there can be any point in concerning oneself with independent reality.

Whether we rejoice in it or deplore it, it is impossible in this area to avoid considerations which fall outside the narrow framework of pure science and can therefore be called philosophical. One such consideration finds full justification in the difficulties in which the ambitious approach of mathematical realism has been entangled for a half century or more. It is at least very dangerous to presume to identify science, that is to say, the description of all experience, with a description of the 'real in itself as it truly is'. As we have noted already, though that might have been the idea Laplace had in mind, we must appreciate that the hypotheses on which it was based have been shown to be contrary to the facts; nowadays, even if the idea is still not absurd, we must accept that all things considered it does appear somewhat too much of a chimera.

We shall not rehearse the reasons, sufficiently explained above, that on the other hand make it very difficult to accept, indeed render incoherent, the thesis of a radical instrumentalism so focussed on human beings as to deny any kind of reality prior to them. Putting both things together we should then recognise that the (most likely) falsity of the idea of Laplace just mentioned is logically equivalent to the assertion that the 'real in itself as it truly is' (a notion that we found to be meaningful) is very likely not reducible to – more exactly, is almost certainly not isomorphic to – the ensemble of assertions based on collective experience, the totality of which constitutes science.

This is an extremely important conclusion. The fact acknowledged earlier, that experience is the basis of science, certainly justifies the principle that science should consider only those assertions that refer to facts and to tests of an experimental kind. But the fact that the totality of independent reality is not isomorphic to what can be described by science (whether because there is too much of it, or because it is something quite different in kind, with only partial correspondences) has the following consequences. It means that if somebody does not want to limit his horizons to the set of what has to do with observations and experiments, and which science therefore can describe, if he aims to look further and be concerned with the whole of reality, his ambition or project cannot *a priori* be deemed incoherent and therefore illegitimate. Somewhat pictorially, but without exaggeration, we could say that this conclusion opens a window in the enclosure within which many minds have unwittingly locked themselves away for so long.

This 'window' – I mean its existence – is so important because of its truly surprising nature, which clearly is that of an opening that is made by rational means – based indeed on today's scientific facts – and that nevertheless leads to something lying *beyond* the totality of experience while not being just an empty yonder. But the great significance of this acquisition should not hide from us the fact that the window involved is very small. At the present stage of our enquiry, it is by no means certain that through it we shall be able to look upon a landscape the features of which we can discern. So we must continue our enquiry.

2 Knowledge of a negative kind

When we are led, as here, by sound arguments to the compound belief both that independent reality 'exists' (more precisely: that it makes sense to see it as existing), and that our science – and more generally our intellect – cannot in all probability provide us with complete knowledge of it, we soon come to consider the traditional 'great problems' in a new light. And first of all, since our capacity for devising a positive account of what exists in reality is limited, we must consider with genuine interest any information that establishes with certainty that ideas previously held to be true or at least more or less probable are in fact wrong or too naive.

Of course the importance of such 'negative information' was clear to thinkers of all times. Indeed, although every age has had its creators of myths, of religions and of philosophical systems – that is to say, individuals who aimed to make positive contributions to what human beings think they know – most numerous among the individuals who aspired to a degree of intellectual rigour were always those who exercised their intelligence on above all the *criticism* of these same myths, religions and philosophical systems, thus aiming at providing essentially negative contributions to these matters. This is understandable and comes to appear quite normal once we note that, as present-day fallibilists have underlined, refutation can in general be successfully performed on the basis of much less information than is needed to establish that an item of knowledge is positively true.

However, in the current situation a new element is present. Whereas until quite recently the beliefs that science was normally concerned to refute were those that came from non-scientific domains, here on the contrary we can observe the refutation (based on information also of a scientific nature) of a doctrine, 'scientific materialism', which by a plainly unjustified extrapolation was once deemed by some to be derivable from science itself. In fact, although the association of the two words 'scientific' and 'materialism' was always judged too bold by some philosophers, it did, though, have some legitimacy at the time of 'classical physics', for at that time physicists thought they could define matter as the ensemble of all atoms *plus* fields, and deemed themselves capable of formulating their science without any reference – not even implicit – to the states of consciousness (or otherwise said, to the sense impressions) of scientific observers. In consequence, the thinkers of that time were justified in their conjecture that 'matter' thus defined is the only irreducible reality. Nowadays though, things are clearly quite different. It is true that biologists, to some degree, manage to explain life in terms of physics alone. But at the same time the principles of that physics have changed so much that fundamentally (and again, short of making an arbitrary choice of one theory from among the set of those, all bristling with difficulties, that aim at recovering strong objectivity) they can no longer be formulated without some reference (admittedly often implicit) to sense impressions and hence to the notion of human mind. The fact that biology is principally concerned with large molecules, for which the classical laws and concepts (easier to handle that their quantum counterparts) in practice provide quite adequate approximations, does not affect the fundamental significance of this epistemological change, though it may for a time obscure it in some respects. The truth is that in so far as materialism claims to be more than just a philosophically irrelevant *methodology*; in so far, that is to say, that it constitutes a particular form of objectivism, be it micro-objectivism or macro-objectivism, we have seen that it cannot be considered as a scientific conception (that is to say, a conception that science supports).

The significance of this item of negative information is worth examining. It may be compared with another item, equally negative, stemming from most recent physics as reported above: namely the fact that what we apprehend as the beautiful simplicity of 'rational' mechanics in no way implies that the phenomena about us are simple. Now the recent realisation, in clear and quantitative detail, of just how complex these phenomena are has not caused any kind of disenchantment. Quite the contrary: many have felt it as a beneficial release from perniciously oversimplified conceptions of the laws that govern the world and of the possibilities for effectively applying them (oversimplifications which had, among others, the grave fault of persuading too many politicians and economists of various countries that it is humanly possible to govern everything dictatorially from on high). Similarly, there is no valid reason why the negative knowledge we are discussing in this section should be felt by anyone as a cause of disenchantment. As in the previous case, it should on the contrary be seen as removing an intellectual block, and so opening up new horizons. Let me explain this point.

In any field, when a particular conception of things has acquired a very high degree of plausibility (because it combines a degree of simplicity with considerable explanatory power over things then known), it goes without saying that possible conceptions incompatible with it lose much of their credibility. It is hard to see how an unprejudiced, objective mind should find credibility in a way of seeing things which explains little if there is another which explains much more. But correspondingly, if new phenomena come to light which the latter cannot explain, and if for that reason the plausible becomes implausible, then the other possible conceptions, which the objective minds previously left to the dreamers, become once more *a priori* worthy of consideration. Correlatively the objective mind feels then more inclined to seek new ideas that, though not being as precise as the current ones, would offer some gains in scope.

This is why, in this field, the objective mind that is conscious of the serious deficiencies of both scientific materialism and of (strict) positivism will be more inclined than most to find out whether there are not other conceptions that are better able to reconcile human beings with themselves than either of those two approaches.

Within a somewhat different order of ideas – though still relevant to the importance of negative knowledge – we should note finally that the obviously merely weak objectivity of physics makes it necessary to place in some perspective the meaning of recent theories about the birth and the ultimate fate (thermal death?) of the Universe. It is interesting to observe in this connection that even the 'strongly objective' models of physics (whose existence was reported earlier) involve such relativising because of the 'distance' which separates their 'basic realities' from the ensemble of phenomena. Thus for example in David Bohm's most recent conceptions the celebrated 'Big Bang', the origin of the phenomenal Universe, appears as 'no more than a small wrinkle' of the real (see [31], p. 192). We therefore need to exercise caution about the philosophical conclusions that many authors claim to draw from the cosmological theories in question. In general these conclusions are false, because they stem from a confusion between different orders of reality.

3 The concert and the record

While independent reality refuses to tell us what it is – or what it is like – it at least condescends to let us know, to some extent, what it is not. It does not conform to the classical schemes of mechanics, of atomistic materialism, or of objectivist realism – in short, to any variant of 'near realism' (it could do so only if it were *totally* unconnected to any imaginable experience, in which case of course the hypothesis that it does would be quite arbitrary). It is thus legitimate to describe it as distant. Moreover, it seems, we may recall, more or less chimerical to hope ever to construct a *scientifically exact* (implying the absence of all arbitrariness) model of it using concepts borrowed from mathematics (as Einstein hoped). Consequently it seems proper to describe it either as 'unknowable' or as 'veiled'. Of these two terms, the second appears the more correct. We remarked above that if independent reality were absolutely unknowable it could not even suggest anything to us, so that on that hypothesis the only concepts of which we could have any notion – and which could therefore figure in our scientific apparatus – would be, in conformity to the views of traditional Kantianism, those the set of which constitutes the *a priori* of our minds. Now we know that the evolution of physics since Kant has invalidated that conclusion. This lends plausibility to the idea that independent reality, though not knowable in the usual sense of the term (a sense which we have seen implies the possibility of precise, exhaustive knowledge) is not absolutely unknowable; once more, it is *veiled*.

This notion, for me the central one, of veiled reality can perhaps best be grasped by means of an analogy inspired by an idea of Bertrand Russell, in which independent reality is compared to a musical concert while empirical reality – the ensemble of phenomena – is compared to a recording of the concert on, say, a disc. Obviously the pattern of the disc is not totally independent of the structure of the concert, but obviously too the recording, consisting of a spatial arrangement in the form of minute hills and hollows in grooves, cannot be identified purely and simply with the concert, which is arranged in time. It would be clearly absurd to suppose that concert and disc constitute one and the same thing. Besides, a Martian who landed on Earth and discovered the disc would never, by studying its detailed spatial structure, be able to reconstitute the concert, whatever abilities he might be endowed with. Is that to say that studying it would give him no ideas about the concert whatsoever? Clearly that would be wrong, since he can indeed get to know the abstract structure of the concert, in a quite quantitative way too. If he is imaginative, and if he is endowed with hearing, he may be able to guess that the hills and hollows he is studying owe their origin to the emission of sounds. Perhaps he could go as far as to picture to himself in some detail how this was done. But if he sets off on that path he cannot but be aware of the inevitable arbitrariness inherent in his proceedings.

Our relation to independent reality appears to be a little like that of the Martian to the concert. Just as he can both grasp and appreciate in genuine fashion the essentials of it, so it seems to me can we discern and appreciate in a non-illusory way (it is in this respect that my return to Platonism is qualified) some very significant features of the real. However to say that we know it (in the exhaustive sense of knowing) would be, once again, mistaken.

We must now deal once and for all with a possible objection which the last few pages have merely touched upon. If in order to keep the line of argument clear of possible sources of vagueness we eschew all analogies, it seems *a priori* legitimate to wonder whether there is not some logical contradiction in the idea of knowing – be it partially, allegorically or in some veiled way – a reality which *ex hypothesi* goes beyond the whole of experience. We have already had to face the question in earlier chapters in connection with the very concept of the existence of such a reality. The answer given there, it will be recalled, was that despite the fact that concepts such as reality, existence and so on are clearly occasioned by experience (in the same way as are all other concepts), it is none the less true that on analysis it turns out that to ban all uses of them that do not conform to a strictly operational code is scarcely coherent. This conclusion was extended, in that discussion, to another concept of the same kind, that of cause as explanation. It is worth noting that the argument on which I based my account is not unlike one developed independently by some philosophers (notably E. Agazzi [51]) and which rests on the idea of distinguishing between the meaning and the reference of a concept. Operationalism does not draw such a distinction. Rather, it requires that the meaning of a concept, that is to say the content of the intellectual representation which it purports to express and which its definition renders explicit, should in no way go beyond its referent, that is to say the ensemble of given facts to which it has been verified that the concept applies and from some of which the concept was devised. But as detailed analysis has already made apparent, when this requirement is rigorously imposed on all concepts, including the most general ones, it leads to operational definitions which are in some cases too restrictive, forcing the concepts into narrow moulds which tend to make them less intelligible rather than more so, as the definitions were originally supposed to do. For these reasons the distinction in question, while it needs to be used with prudence (indeed, in the technical areas of microphysics, not at all, for there operationalism is a reliable guide), seems on the whole to be justified, at least when we are dealing with fundamental concepts such as reality, causality in general and so on.

Thus removing this objection leads to the conclusion that there is some legitimacy to the philosopher's indulging in a little metaphysics. Clearly though, the dangers inherent in all 'mixing of modes' have to be borne closely in mind. While a physicist such as myself may admit that such an undertaking is legitimate, this does not mean that with respect to certainty or their degrees of 'rational conviction' he would put metaphysics on the same footing as physics. If the analogy of the concert is of use it is precisely to show that this is not the case. Under the circumstances it might seem surprising that I am here willing to accept the validity of an activity which, gauged by the yardstick of science, seems so highly conjectural. I think however that in this matter we must take a rather wider view of things than most scientists worthy of respect instinctively tend to take. They have got into the habit – in consequence of a long history of strict testing of theories, often leading to their rejection – of identifying what is rational with what is certain or at least with what is subject to the 'rational conviction' that (whatever may have been said of it) is a dominant feature of the sciences. In their eyes, all the rest is irrational. Now to me such a position seems too sharply drawn and too *a priori*. It does not seem incoherent to me to admit the possibility of rational activity that does not issue in 'demonstrative certitude' in the sense that we scientists use the expression. The only condition that I think we can nowadays impose on those who engage in such activity is that they do not claim for their findings a degree of rational conviction greater than what they actually have.[3] The restrictive nature of such a condition was – or seemed to be – less marked in the great ages of metaphysics because there was none of the element of comparison so evident today. But once again, in my view that must not lead us to deny the possibility *a priori* of any rational metaphysical activity (though it goes without saying that such a conclusion does not require that we immediately take sides as to what might be the results of such activity).

4 Life, consciousness, cause and time

There are, as we know, no perfect analogies. Above, that of the concert proved illuminating, but on one important point it is deficient. It might, in effect, lead us to believe that independent and empirical reality are ontologically to be put on the same footing. Plainly they are not. Indeed, given that empirical reality is by definition the ensemble of those phenomena that human beings can become acquainted with, the question of its accord with independent reality becomes essentially that of the relation between human consciousness and such reality.

In its turn, this question cannot but give rise to that of the relation between life, with which consciousness has ties, and independent reality. Must we put life rather 'on the side' of independent reality, or on the side of empirical reality?

Given the great progress made by present-day biology, there is scant cause for much hesitation. Nowadays life seems to us to be in essence a particular ensemble of phenomena which operate at the macromolecular level. The complexity of these phenomena certainly strikes us as marvellous. So too the emergence of order that accompanies them. And taking into account the immense difficulty of carrying out scientific investigations that relate the main parameters (average time taken for the evolution of organs and species, and so on) to the facts of physics, we seem to be in no position at present to make any hard and fast claim as to whether life is or is not reducible to physics. Nonetheless, it is true on the one hand that currently there is no convincing argument that it is not, while on the other, whether we deplore it or not, there are two facts of a very general kind that tend to support the 'reductionist' thesis. The first is that simple equations may have highly complex solutions (and similarly that quite ordinary groups of transformations have very complicated representations). The second is the emergence of order from disorder, which as is well known is characteristic of certain phenomena that are at the same time open and out of equilibrium. Undeniably, the balance is in favour of a moderately reductionist thesis; the characteristic features of life seem to be capable of inclusion without fundamental conceptual difficulties among the grand ensemble of that totality of phenomena we have agreed to call empirical reality.

Does the same apply to consciousness? The fact that we do not know with certainty of any 'consciousness' that is not associated with living things must of course play an important part in any reflection on the problem. It certainly seems to show that, as we have already said, consciousness follows in some way, via the intermediary of life, from the phenomena of physics. Here though we should beware – as I intend to explain – of a natural chain of thought that arises from the habit we may have acquired, particularly if our scientific education has been elementary, of conceiving everything in mechanistic terms. It is certainly true that the complicated arrangement of the macromolecules which make up living cells can 'in the main' be described in such terms. This is because the more the dimensions of a physical system approach the macroscopic scale, the greater the degree to which the laws of classical physics – and hence mechanistic descriptions – constitute normally acceptable approximations to the laws which truly govern the system, which of course are quantum laws. The macromolecules involved here are, as we have noted, generally[4] large enough for this kind of approximation, which has the advantage of being relatively simple, to be perfectly adequate for the kind of problems to be dealt with. It is therefore quite understandable that we are usually satisfied with descriptions of biological phenomena that are expressed in terms of classical laws. On the other hand, such laws can as we know be stated in terms of strong objectivity; if that were true at all scales, it would be possible to give them the status of descriptions of reality in itself, and it would be legitimate – and not at all unreasonable – to consider the physical objects to which they refer as existing in themselves.

The natural chain of thought I said we must guard against consists precisely in acting as if these conditions were met – which they are not – and hence considering – mistakenly – that in any entity, in particular any living thing, the molecules, the atoms and their constituent parts are truly objects of which it can be said that they and their properties have an existence that is quite independent of anything connected in any way with the conditions under which the human mind can know about them. If we give way to this chain of thought then clearly it will become normal to consider consciousness as a simple emanation from some or other of these molecules or aggregates composed of them.[5] If on the other hand we bear in mind that giving way to it is incoherent, such a way of seeing things itself becomes illogical. Indeed, since objects in general, and the molecules of the brain in particular, are phenomena, since phenomena are essentially – in a theory whose objectivity is only weak – 'appearances that are valid for all', and since finally the notion of an appearance obviously is meaningful only if the concept of a state of consciousness is first posited, then clearly the merely weak objectivty of the true (quantum) laws tends to reverse the order of subordination. Instead of the existence of consciousness following from the existence of objects, the existence of objects now seems somehow to follow from the prior existence of consciousness (or consciousnesses).

To define any notion – in particular the notion of consciousness – by referring to the notion itself is to get caught up in a vicious circle. That is why, once more, the concept of consciousness is necessarily prior to the concept of phenomena, since the latter is defined (as we have seen it must be, given the weak objectivity of the physical description of the world) in terms of appearances valid for all, that is to say, in terms of facts of consciousness. But it is also obvious that, whatever its significance may be, the conclusion we have just come to in no way negates the obvious fact we noted at the outset, namely that consciousness seems always to be associated with an empirical reality that is material. How to reconcile these two facts, which at first glance seem to rule out one another? It is certainly difficult, but we should not say that it is impossible. As I stated elsewhere some time ago, my own view is that it can be done by conceiving thought and empirical reality as two complementary poles which give rise to each other within the realm of independent reality. Now there is a measure of obscurity – scarcely surprising in such matters – in such a thesis. But we may perhaps grasp something of the sense of it by making further use of the method of analogy. The reciprocal generation at issue may, without pressing the comparison too far, be likened to the way in which language and responsiveness also give rise to each other within the realm of all human consciousness.

We may wonder whether it is possible without further ado to assimilate the conception just suggested to one that is more common in a certain widely-read branch of the literature, according to which matter and consciousness are held to be two realities in themselves, which are capable of mutual interaction. Certainly the two conceptions have certain things in common – but it is clear that they cannot be identified. In the domain of physics the only serious arguments which seem to favour the second are, as we have seen, those propounded most notably by Wigner and known as 'the paradox of Wigner's friend'. These make reference to the reduction of the wave packet at the time of a measurement event; I have discussed them elsewhere [2B] and have acknowledged their force. At the same time I have also noted their weaknesses. In outline, these amount to the problem that if one truly considers the reduction of the wave packet as the action of one reality in itself (the consciousness of the observer) on another reality in itself (the matter of the system under observation), then it seems impossible to confine such effects within acceptable limits. As we saw in Chapter 10, we must resign ourselves to holding that such and such a particular observation of the Moon may confer on the Moon a definite location that previously it did not have. Any conception which appears more 'reasonable' in that the effects of the registering in consciousness are something more subtle and less dramatic cannot readily be fitted into an explanatory scheme that one would try to base on the reduction of the wave packet and a dualist ontology (here, the two realities in themselves). This is one of the main reasons which now lead me to draw a clear distinction between such a conception and that of veiled reality explained earlier, and to prefer the latter.

In accord with the classical tradition in philosophy, I earlier gave to consciousness that is both generative and generated the fine name of 'thought'. But are we concerned only with reflexive thought? With thought that 'knows that it is thinking'? In my judgement, the conception of veiled reality does not lead to such a restriction. It is not hard to see how such a restricted conception of thought came into vogue. In an age of rather triumphalist 'scientific materialism', it was a matter of trying to give a definite form to an intuition of which it was hoped that at least it would preserve *one* domain that the dominant philosophy could not violate. But on the one hand the method then chosen now seems of doubtful efficacy, since nowadays it is no longer clear to us why reflexivity should constitute an insurmountable barrier for artificial intelligence, while on the other scientific materialism itself is now in such an uncomfortable position that as a way of conceiving the world it no longer represents a genuine danger of error. (Alas the same cannot be said of the 'materialism of daily life', the narrowing of the intellectual horizons of the majority and the inability of the education system to enlarge them, which does indeed represent a serious danger for highly developed societies; but that kind of materialism has nothing in common with the other apart from the name.) Indeed, consciousness as I think of it cannot in any way, for straightforward reasons of logical coherence, be reduced to a mere property of matter: for it is consciousness which in a sense carves out the atoms within the body of reality. But in essence it does this by means of operations and 'registering in consciousness' (such as deploying an instrument or observing a signal) which have about them nothing that is specifically 'reflexive'. Animals could thus quite well be channels of such 'thought' too.[6]

Reflection on life and on thought inevitably leads to reflection on time; and in its turn, time cannot be separated from questions concerning the idea of cause.

Neither of these problems is easy – that is the least one can say – but the second is perhaps even more complicated that the first. However, that is not to say that philosophers and scientists have not both made great efforts to clarify it. Many of the former have proposed the maxim that any cause must belong to the same category as its effects, eliminating in particular thereby any metaphysical cause of physical effects. This way of looking at the matter has been discussed in some detail in an earlier chapter. Certainly anyone is free to take it as a restrictive definition of what he is willing to call a cause. But then a need for explanation remains, in the present state of physics, unsatisfied. This is why reference to an independent reality (even though it now proves to be veiled), remains necessary today. That one should call it an 'explanation' only, and refuse it the name 'cause' I readily consent to, although this is a choice which to me seems neither highly significant nor particularly convenient.

Since the time of Galileo, scientists have for their part let the notion of final cause bear the brunt of their criticism, and have, it would seem, managed to persuade the 'laymen' that the notion is totally inane. I am not absolutely certain that in doing so they have been altogether well-advised.

As a preliminary let us first note that in the most substantive works on advanced modern physics (such as relativistic quantum physics and quantum field theory) formulations of the material may be found that make it rest on extremum principles which have at least as much (and in fact much more) in common with descriptions in terms of final causes as with statements of the traditional mechanistic kind. Nevertheless, there *is*, admittedly, just one domain in which the limitation to anterior causes – and the exclusion of final causes – remains an essential element of scientific description. It is that of cause in the restricted – and technical – sense of the term. 'Phenomena', as we have seen, include an important component contributed by human beings, in the sense that it is our perceptual and intellectual faculties which in large measure demarcate these phenomena within the body of the real. In consequence, the classical philosophical problem – given two phenomena linked by a causal relation, which is the cause and which is the effect? – can always in practice be settled otherwise than by appeal to the simple definition, equally classical, that 'the cause is the one that is antecedent to the other'. The cause, in reality, is the phenomenon which is directly dependent on us, or of which by an act of imagination we can conceive most easily that it could depend directly on us; and it turns out – in outline, this is then a *law* and not a definition – that the cause-phenomenon is never posterior to the effect-phenomenon. Since this order of succession is a law and not a simple convention, it can have implications that are objective (that is to say, that are not themselves mere conventions) without there being anything paradoxical about the fact. This is indeed the case. An example the physicists are well acquainted with is provided by the so-called 'dispersion' relationships, which connect the real and imaginary parts of the amplitudes of certain phenomena caused by a collision or the closing of a circuit. These relationships – which are experimentally verified – would be false if final causes existed in the sense being considered here. In that sense, the relations between cause-phenomena and effect-phenomena certainly do seem to exclude, otherwise than by convention, the notion of a final cause, which is what scientists have always maintained.

These arguments, excellent and beyond reproach as long as they are deployed within the empirical domain, cannot be extrapolated to that of causes lying 'within' independent reality and taking effect in phenomena. While therefore the idea of rigorously excluding the notion of final causes could in a sense be considered scientifically justified in the time of classical physics – for then the distinction between independent reality and empirical reality did not seem necessary – today there is no longer any justification for such a claim.

This last conclusion – very important if indeed distinctly meagre – of course does not at all mean that we *must* consider final causes. Nor does it imply that such a notion is altogether clear (if independent reality is not in space-time, it is quite obvious that it is not clear). But it does deserve sympathetic attention from people and groups of people who are aware of the fact that Aristotelianism and Thomism are still repositories of genuine emotional values, so that any removal of the refutations that claimed strictly to rule them out of court can open up new horizons of thought. Now this is precisely what the effect could be of acknowledging that there is nothing intrinsically absurd in the notion of a final cause as understood here. We know that the God of Aristotle is in essence a final cause. And if there is some way by which the notion of a God of love, so essential to Christianity, can be reconciled with our experience of, for example, natural catastrophes, no doubt it is primarily in the observation that love acts in the manner of a final cause that we ultimately have to seek it.[7] Perhaps it is not exaggerating too boldly to dream of similar conceptions taking shape in the future (for example, what has been said here opens the possibility that the existence of what I have termed elsewhere [46] 'calls of Being' is not wholly illusory).[8]

If such conceptions come to see the light of day, it could be that they will have an effect on artistic and musical creation as beneficial as that which Christian thought had during the High Middle Ages and the seventeenth and eighteenth centuries. Artistic creativity (which I take to include poetry, music and some kinds of prose), I would maintain, cannot reach the greatest heights unless it is inspired by some disinterested hope of exalting Being in the form of a being – Ammon, Christ, Nature, or some other – that totally transcends the artist and yet, as if in a dream, is infinitely close to him. I would also maintain that apart from the ashes, still warm, of splendid but short-lived cultural bonfires, the destruction of such hopes, which was much furthered in the course of our century, left scarcely anything but a wasteland: television variety shows, juke boxes . . . . And, it must be granted, also some deserving cultural efforts, but which for all the enthusiasm and talent of their promoters are hard put to rise above the general level of the wasteland. I would therefore like to see a rebirth of a 'sense of Being' capable of infusing with new life the fresh, candid and elusive hope I mentioned. The germ of such a rebirth, it seems to me, might lie in the fact that we now have a solidly-based idea of an independent reality that is neither totally inaccessible nor totally reducible to trivial notions.

To be sure, there is a difficulty here. It lies in the want of *proximity* of such reality, a want that seems to follow from its nature as something non-banal – albeit that trivality and proximity are not totally synonymous. No doubt it is such a need for 'proximity' that has drawn philosophers to conceive of what they call 'an immediate apprehension of the thing itself', and to see in such an apprehension what is ultimately the truest communication human beings can have with the real. What are we to think of such an idea? This, I suggest: it is ambiguous and heavy with meanings which, though not necessarily illusory, are erroneous *if* we take the idea as baldly stating that the object we thus apprehend, a physical object localised, singular and unique, truly constitutes an element reality in itself – which latter would then be no more than a mere juxtaposition of objects linked by forces. Whether we deplore it or not, such a return to near realism can only be accepted by minds ignorant of physics. However, looking at it from another angle, how could we remain insensible to the impression of direct communication with the localised, singular object evoked by the philosophers concerned, and from which artists, poets and composers draw their finest inspirations – ultimately the most precious part of our lives? Are we obliged by our other knowledge to range such impressions under a rubric similar to that to which we allocate optical illusions? By no means. It is obvious – though nevertheless it is worth stating – that the hypothesis according to which some particular impressions of this kind are *indications* of truths about reality in itself cannot be refuted.

Is it possible to go further, to be more precise? Yes indeed, if we are prepared to go in for invention (and in such a domain invention is very easy[9]), but alas no, if we reject the arbitrary. So at this stage a certain conceptual asceticism is necessary, even if it feels frustrating. The need for it illustrates once more how immensely difficult it is to reconcile the best-founded conceptions of the two hemispheres of the brain in any detailed but at the same time not superficial or arbitrary manner. Let us therefore stay with the impalpable notion of a 'reality that lies behind the things'. It has the unusual virtue of being both charming and reasonable. Does this not give it some value?[10]

The notion of cause is difficult to separate from that of time, and time is at the heart of all that is important to human beings. It is fitting therefore to draw this book to a close with some ideas relating its content with the notion of time.

In outline, we seem to have encountered or entertained three kinds of time in the course of the preceeding investigations.

The first of these is what philosophers often call 'physical' or 'mathematical' time. It is the variable t which figures in the mathematical expression of the elementary laws of physics. Since relativity appeared on the scene, this variable has been associated with space as the fourth component of what is called 'space-time'. The invariance of the fundamental laws of physics when the direction of time is reversed may, for all that is significant here, be considered as a general property of these laws. If physics could be reduced to a small number of such fundamental laws, which could themselves be expressed in terms of strong objectivity – if, that is to say, physical realism could be considered a plausible vision of things – the 'reversible' time we are talking about would have a fundamental role. In accordance with the Einsteinian conception, it – or more exactly the space-time of which it is a component – would in this domain appear as the 'true reality', while the other kinds of time to be discussed below would on the contrary appear as illusions or metaphysical dreams. As we have seen however, that is not the case. Not only does it seem necessary to give up the conception of physical realism, which unified the two notions of reality in itself and of empirical reality, but also it seems probable that reality in itself is prior to space-time. At any rate, the tension between quantum physics and the notion of locality makes the idea that effectively this is so seem plausible. Under the circumstances, then, we must recognise that although the reversible time of mathematical physics remains a quite essential element in a set of immensely effective calculation rules, its 'ontological status' appears nonetheless precarious. It is not an element of independent reality, and as an element of empirical reality it has to give way to irreversible time (to be discussed next), at least in respect of most of the phenomena we normally refer to by that term.

The second kind of time, irreversible time, rules our common experience. As we have seen, it pertains to the world of phenomena, or in other words, to empirical reality. As with all that reality, it is not of such a nature that we can declare it a component, or an element of a description, of independent reality. At any rate, some very serious *a posteriori* arguments, laboriously but soundly and rigorously derived from contemporary scientific data, go against such an assimilation, whether we like it or not. However, such time, that of life, death, the development of clouds and of practically all the phenomena we perceive, has for that very reason a 'density of reality' which cannot be denied. By contrast with the mathematical time just discussed, there is *at least one* kind of reality, empirical reality, into which this second kind of time fits very well. So it is not absurd to speak (albeit loosely) of it as being 'more real' than the first kind. And this, after a fashion, provides reason after the fact for those philosophers who, sometimes implicitly and not fully aware of what they are doing, identify 'true time' with the consciousness human beings have of time.

Are these two the only notions of time we need consider? It seems to me that the reflections of the last few pages have enabled us to glimpse the possibility of a third conception of 'time'. Unfortunately, 'glimpse' is exactly the right word. To put it another way, anyone who tries to evoke this third kind of time cannot, even if it goes against the grain of his personality, avoid resorting to modes of language somewhat reminiscent of the discourse of hermeneutics, or reminiscent, at least and unfortunately, of the way in which such discourse cautiously holds on to the edge of problems while expatiating on their profundity, and hence their obscurity. Thus we all know of Heidegger's pronouncement 'man is the shepherd of Being', a true *parable*, deliberately enigmatic, profound – who knows? – and perhaps interpretable in terms of veiled reality. We are perhaps less familiar with another 'parable' of his, 'time is the horizon of Being', which Beaufret [54] 'explains' as saying that 'time . . . serves as heralding Being in the existent'. Anyone who tries to elucidate in what this third kind of time, barely glimpsed, consists will to some extent be obliged to have recourse to such 'invocations', which both speak so appealingly to the right hemisphere *and* seem to convey something to the left hemisphere of our brains, albeit that the latter cannot always say exactly what *is* conveyed . . . .

In the present case however, there are certain elements which narrow down the notion we are seeking, elements which though certainly partial and far from fully meeting the expectations of minds dedicated to lucidity are nonetheless not devoid of meaning and can be described. This third 'time' I am thinking of – fundamental time – is a dimension of veiled reality, or as we could also say, of Being. It is a dimension which, it goes without saying, cannot be conceived as temporal in the narrow sense that implicit reference to one of the other two kinds of time would necessarily give to such an adjective. But it is a dimension which seems truly to exist, and which we may conjecture is not *entirely* hidden from us. The real, I have said, is veiled. It is for us what the concert is for the Martian who has only a recording of it. But it could well be that unlike the concert in that fable, veiled reality is not of such a nature that it can be glimpsed in only one way. Without doubt, as we have said, the reasoned paths signalled for us by science provide one such way, for even if scientific models and laws do not describe it 'as it is' it remains probable, given the arguments that have been presented, that they reflect something of its structure. But the existence of man overflows the framework of science on all sides. In particular, in human experience of time not only is there irreversible empirical time, which scientific thought is now beginning truly to know how to describe by associating it with elaborate mathematical entities. There is also another kind of experience, with an essentially emotional colour to it, which the philosophers we mentioned, along with many others, have devoted much attention to. If, as I have strong grounds for believing, the first kind of experience, that of phenomena (and of irreversible time in particular) in the sense of 'facts registered by a cool eye', is not totally misleading but rather gives us some *insight* into the structure of reality, on what arguments could we base the hypothesis that the same is not true for the other? For my part I find plausible the idea that, when taken literally, our apprehension of time, just as that of the absolute nature and the singularity of any object, is an illusion, but that if they are taken 'at one remove' none is totally illusory. It is conceivable that, on the contrary, they should be fleeting but precious reflections of the 'truly real'.

That we are here moving in the realms of pure conjecture, indeed of thick fog, I would be the first to admit. But the conjecturing is slightly less hazardous and the fog not quite as thick as is the case with some purely philosophical queries which however presume – quite rightly – to the title of 'thoughts'. So let us content ourselves with having arrived at the point of confirming the rational possibility that there is a kind of experienced time which, far from being a mere ordering of the phenomena, has something to do both with man and with Being, and thus constitutes a kind of bridge between them.

APPENDIX I

On the 'degree of certainty' of empirical claims

Is there any such thing as a purely observational statement, one without any admixture of theory? Does the idea of direct observation make sense? Properly speaking, do observational terms exist? Similarly, are there statements – at least statements about particular events or circumstances – of which, given the observed facts, one can say with complete certainty that they are true? These various questions cannot be reduced to one, but nowadays they do bear closely on one another. Without of course seeking to be exhaustive, we should examine them here in order to fill out the suggestions made in Chapter 2 on observational terms.

Let us begin with the first of these questions. Are there truly two distinct classes of observations, those that are indirect, in the sense that they involve instruments, and those that by contrast are made directly? Rightly, the answer of the great majority of philosophers would be no. Consider for example the light emitted or reflected by a body: they would point out that it would be possible, if one so desired, to detect this light using appropriate instruments which when placed in the vicinity of the body would convert the light into an electric current, which could then be analysed by a computer. By definition, observing the object in that way would be indirect. But it is obvious that this would still be so if the apparatus in question were replaced, first by a microscope and a projector screen, then by a simple lens and the same screen. Now then, what difference does it make if the lens is the crystalline lens of an eye and the screen is its retina? The physicists and the physiologists would agree that qualitatively there is no difference – from which we must conclude that what we call taking note of an object by 'direct vision' is qualitatively no different from what we call examining it 'indirectly' by means of instruments.

This line of reasoning is quite sound. But there are a number of contemporary authors who, having completed it, then gallop through the next stage. This, if we pause to analyse it into its components, proceeds as follows. Any observation carried out by means of an instrument rests in part on theory, namely the theory of how instruments of that type work. Now any theoretical development includes generalisations and extrapolations, and therefore possibilities for subsequent correction (by a later theory). Given this fact, any indirect observation is 'laden with theory' and hence uncertain. And since, as we have seen, there is no qualitative difference between direct and indirect observations, ultimately we must extend that judgement to cover any observation whatsoever.

In philosophy of science as it is now taught, such a conclusion is, often enough, elevated to the status of a fundamental truth. It is usually illustrated by examples drawn from the history of science and having to do with erroneous conclusions drawn from diverse observations. This kind of trend in the teaching of the philosophy of science has thus given to most – if not all – student apprentices in philosophy the idea that there is a certain want of sophistication about men of science in this respect, which results in them taking experimental data too much for granted. Such naivety is said to be gross, since it is due to a misunderstanding of extremely simple arguments and established historical facts. Philosophers of today and tomorrow are said to be delivered from it for the very reason that they keep these arguments and facts in mind.

In any argument, the greatest naivety is not always where it seems to be. In the present instance, even though I do not want to pass summary judgement on matters of substance, I do feel obliged to insist on the fact that if there is any remaining uncertainty about the observations on which science is based it is to be found at a level of subtlety quite different from that of the above critique.

The critique itself is not valid; the reasoning involved only *appears* to be coherent. To convince ourselves of that, we need only to remark first that to see an object, or to be more precise, to see that the lamp of a signalling light is on rather than off, there is not the slightest need for us to be acquainted with the theory of optics as applied to crystalline lenses, nor with the structure of the retina. Suppose that several people are looking at the lamp at the same time. If they should happen to disagree as to whether it is on or off, the difference between them is statistically never reducible to a difference between those individuals who are acquainted with the theory of optics and the structure of the retina, and those who are not. If there is a difference between the groups it has to do with factual circumstances – good eyesight and the like – which have nothing to do with any such knowledge. Thus although in one sense, as we have seen, all observation is indirect, it is nevertheless true that there are observations of which the result is independent of the theory of the instrument involved (the eye in the example) and hence with its vagaries.

That being so, we must grant, however, that even in the simplest cases intersubjective agreement – agreement among those who are party to the same observation – is never completely assured. It is always possible that there will be the odd person who does not observe the same thing as his fellows. But in any event, and this is the essential point, in simple cases – such as those which involve declaring whether a lamp is on or off – there is always an overwhelming majority (those whom, by definition, we call 'normal in this respect') of the total population among whom intersubjective agreement is reached spontaneously and without problems. At this stage the philosopher can of course, as we noted, retreat to a position of pure logic and maintain quite legitimately that the opinion of the majority, however imposing it might be, can never be elevated to certitude. But he can hardly base his case on the argument from theory-ladenness, which is the one we proposed to invalidate.[1]

But what of the examples, it may be asked. Do not the numerous historical errors alluded to above prove that observation is riddled with uncertainty?

Here again the answer is no. In themselves the examples prove nothing of the kind because the errors to which they refer do not bear on the observation itself but rather on the theory by means of which one sought to explain the observation. What such examples can prove – and prove quite well – is that due to the way theories evolve, observations that at one time were considered as trivial may later come to acquire considerable significance and genuine importance. That this fact is worth noting is undeniable. But it is something quite distinct from the supposed uncertainty inherent in observation itself. The discovery in 1856 of the strange 'Neanderthal skull', we are told, was held to be of scant significance, and the anomalous features it displayed were attributed to its being the skull of an idiot. But forty years later the discovery at Trinil of a very similar skull led to some fruitful developments. The difference between the two cases is that in the meantime Darwin's theory of evolution had appeared on the scene. Certainly the example demonstrates well the essential role of theory in the development of science. But it would nevertheless involve an unacceptable semantic slide to claim from it that the measurements carried out on the first skull at the time of its discovery were riddled with error and uncertainty. . . .

So we see that the idea that there are no purely observational statements is not as easy to justify as many authors think. It would be quite wrong to say that the thesis according to which it is possible to define such statements rigorously has been definitely refuted and rendered unacceptable, and that by elementary considerations. Indeed, it seems that in the minds of the most eminent philosophers the question remains open. Quine, for example, after having called 'incitative sense' of a statement the totality of sensory stimulations that would lead us to consider this statement to be true, goes on to define observational statements as those which are such that their 'incitative sense' (a) remains constant even if the network of laws and concepts is changed as much as possible, and (b) can be said without contradiction to exhaust their meaning (quoted in Hesse, *The Structure of Scientific Inference* [55], p.27). The very fact that Quine provides such a definition seems clearly to imply that he considers the set of all statements defined in such a manner as not being empty, even though elsewhere (in *From a Logical Point of View*) [56] he has no hesitation in declaring that no proposition is quite secure from revision.

Quine's definition, we have noted, puts the emphasis on sensory stimulation, and indeed if we aspire to as great a degree of certainty as possible, it is clearly to that that we must turn our attention. To take a celebrated example of Russell, the statement 'there is a dog here' cannot in any way be said to express a certainty because it is conceivable that the dog is only a mirage, a film image or whatever; whereas by contrast the statement that 'Peter, James and Mary see a coloured, dog-shaped patch' is not subject to the same objection. Believing – but very probably quite mistakenly, as we have noted – that science can legitimately see as its aim the detailed description of reality as such, many philosophers are not satisfied with statements of the second kind, and when they claim to be discussing the certainty of observational statements they therefore concentrate on statements of a realist kind, such as 'there is a dog here' (or 'a molecule' or 'a quark'). They have no difficulty in showing that those statements are vulnerable from the point of view we have been discussing. But they find it more difficult to demonstrate the uncertainty of statements of the 'see a dog shape' or 'see (or not see) a lighted lamp' kind.

Some of them, however, do tackle this latter problem lucidly and furnish arguments that do indeed address the question at issue. This, for example, is the case with Mary Hesse [55], who proceeds by taking very simple words referring to sensory descriptions – the word 'red', for example, or 'simultaneous' as applied to two events – and seeks to show that strictly speaking these words are not directly descriptive and independent of any established law. To do this though she is obliged to invent some extremely odd situations. She imagines, for instance, an isolated tribe afflicted by a particular visual defect such that they cannot distinguish between light green and red, and dark green and black. She points out that if this were so the day the members of this tribe would learn physics they would be constrained to alter their normal use of the words 'red' and 'black', and that the same would apply if the tribe began to have dealings with the rest of the human race. Similarly, she notes that before 1905 the notion of absolute simultaneity between distant events was not questioned, and that therefore at that time the idea of replacing the hypothesis of universal time and the notion of absolute simultaneity with the hypothesis of the constancy of the speed of light (with respect to different frames of reference) would certainly not have been seen as a way of founding theory more directly on experiment but rather just the opposite. She quite rightly uses this example to show that it is impossible to talk sensibly about observations that are more direct or less direct, or of concepts that are to a greater or lesser degree laden with theory, in the absence of a given framework of fixed laws. And, continuing in the same vein, she seeks further to show that even the claim that two events occurring in one and the same limited region of space-time (that in which the observer is located) are simultaneous does not have in all logic the character of direct and absolute certitude which is credited to it. To that end she imagines – what is readily conceivable – that the observer's assessment of the simultaneity or non-simultaneity of two events can depend on the strength of the gravitational field, supposedly highly variable, in the region. In such circumstances, the claim that 'the observer can always judge directly the simultaneity or non-simultaneity of the two events in question' is true, she tells us, only if we interpret it as referring to the direct impression of simultaneity or non-simultaneity experienced by a particular observer. It is not true once we demand – as we must – intersubjectivity, that is to say, a possibility of exchanging information leading to an agreement among several observers.

To Mary Hesse's first objection we must answer, it seems to me, that the tribe she imagines is *ex hypothesi* made up of a sample of those 'deviants' whose existence we recognised above. Their presence does not affect the validity of an intersubjective agreement as long as they remain a quite small minority. If this were not the case, if for example the tribe in question were a fairly large proportion of the whole human race, then things would be simple: the scientific community would stop using red lamps in signal lights – just as it does not build any measuring instrument which involves distinguishing between 'good' and 'bad' smells, a distinction which is too subjective and personal.

To her second (and ingenious) objection it is appropriate to make a similar reply, I think. If it happens that two observers, each equipped with refined instruments for measuring time and each working in the same room cannot arrive at agreement on matters of simultaneity – or better, relative antecedence – of two events occurring there, and if this novel phenomenon was repeated in such a way as to exclude accidental error, what should we do? Clearly we would have to find an explanation, and that imagined by Mary Hesse could ultimately turn out to be acceptable. In the list of experimental conditions which serve to specify the operational definition of relative antecedence it would be necessary to include remarks such as 'in the absence of a gravitational field' or 'for a given gravitational field'. Conversely, if the inclusion of such restrictions sufficed to restore intersubjective agreement, we could then consider that the explanation of the phenomenon by gravitational fields is the correct one.

By thinking up phenomena more and more removed from those dealt with by our physics it is of course possible to prolong that kind of dialectic to infinity. What certainly is true is that each stage of such an imaginary journey would involve scientists in a process of concept splitting that would be long and difficult, mainly because these concepts were formulated in the first place on the basis of limited experience. The plight of the scientists would be the same as that which faced Ancient Egyptians (who, having no river but the Nile, used one and the same concept for the two notions of 'north' and 'downstream') when they discovered the Euphrates. But why should we suppose that in the future this task will prove insurmountable? Human reason has always managed to cope with it quite successfully so far.

APPENDIX 2

Two possible kinds of dependence at a distance, as gauged by the yardstick of intersubjective agreement

Let us consider once again the phenomenon of correlations at a distance, already examined several times in Chapters 5 and 6. Two 'particles of spin ½', U and V, that have interacted in the past and that, owing to this, are in some well determined total-spin state are both subjected to a measurement of their spin component, in direction **a** for the former and **b** for the latter, in such a way that the two measurement events are spatially separated (that is to say, light does not have time to travel from one to the other). Let us again call the results of these measurements α and β respectively (α, β = ± 1, in appropriate units). As we have seen, if λ designates the objective state of the system $U + V$ when it emerges from the source and p is the probability that if λ is fixed, then a definite value of α, for example 1, will be observed, the hypothesis that p depends neither on **b** nor β – the hypothesis which in essence constitutes the 'principle of separability' – entails inequalities that are violated (Bell's theorem); accordingly, the hypothesis cannot be maintained. This then amounts to non-separability.

It can be said that this established violation of the hypothesis makes manifest the existence of an action at a distance that is instantaneous in a certain frame of reference. In other words, it can be said that it constitutes an exception to the theory of relativity interpreted in a realist sense. But as we noted in the main text, given that the λ are not observable this violation does not necessarily imply a refutation of the same theory interpreted in an operational sense. In fact, as also noted, it is sufficient that the quantum prediction rules (the quantum axioms in their usual formulation) be true for it to be impossible for any signal to be transmitted over a distance at greater than the speed of light.

That much having been said, it is clear that *a priori* we can imagine three ways in which a theory can violate the principle of separability. The theory might hold that p depend on β but not on **b**; that p depends on **b** but not on β; or that p depends on both. Since β does not depend directly on the operator (who is indeed free to decide to make the measurement, but not to choose what results will be obtained), it is fairly clear that a theory of the first kind, although it incorporates non-separability, does not (even with λ fixed and supposedly known) allow of the instantaneous transmission of signals by means of the apparatus being considered. Now as several people [57], [58] have recently stressed, if we identify λ with the usual quantum state ('state vector'), quantum mechanics is just such a theory of this first kind. This has led one of these authors, A. Shimony [57], to talk of 'peaceful co-existence' between quantum mechanics and relativity.

By contrast, theories which involve *non-local* hidden variables (the only ones that remain acceptable) are of the second (or more exactly, of the third) kind. Those of these theories that do not exactly replicate all the observable predictions of quantum mechanics are therefore in danger of violating relativity, even at the operational level – which should, it seems, suffice to refute them. It must be emphasised however that this objection cannot be set against the theories involving non-local hidden variables that do exactly replicate these predictions. This follows from what was pointed out in the second paragraph of this appendix. Of those theories too we may legitimately speak of peaceful co-existence.

For reasons elaborated in the main text, I do not think it possible – neither in the future nor *a fortiori* at present – to formulate a truly credible theory of reality in itself. This does not mean that it is impermissible and indeed pointless to try to devise consistent models of such reality, since (see for example Chapter 10) even if none of them is plausible the mere fact that they exist is enough to convince us of the consistency of the general conceptions within which they can be constructed. Hence the above remarks make it possible to distinguish two directions which, *a priori*, such a search for 'realist models' could fruitfully take. One leads in the direction of theories in which the state vector (*alias* the wave function) not only has ontological status but also provides the most detailed description of independent reality that we can conceive. In the main text I have given the reasons that lead me to judge that the difficulties that path bristles with are, when all is said and done, rather dissuasive. The other direction, leading to models that involve non-local hidden variables and do exactly replicate the observable predictions of quantum mechanics, is also beset with great difficulties. However as noted in Chapter 10, it is rather in that direction that nowadays I see some glimmer of light. A general idea which underlies these models is that only the coordinates of particles (or the entities which play an analagous role) are observable, while quantum potentials by contrast are not. Another way of expressing this idea is to say that within a reality composed of entities that are extremely non-local (here the 'quantum potentials'), our consciousness marks out a domain of apprehension in which only *local* entities are to be found. Arguing from the case of the pair of particles $U + V$ considered at the beginning of this appendix, it is possible to show quite rigorously – using formulae that are too long to set out here – that intersubjective agreement not only creates no problems in such a non-local hidden-variable model but is indeed a necessary consequence of it. At first glance this might seem of small importance. For we naturally tend to regard such agreement as something to be taken for granted, given that it is easily explained within the framework of naive realism (that is to say, within a theory which regards sense objects as things in themselves). But as we have seen, it is just such a framework that nowadays we must abandon. And once we do that, intersubjective agreement becomes something that requires explanation. In some – dualistic – theories of the first kind mentioned above, such explanation cannot be other than most peculiar, since it is based on the fact that such agreement is merely apparent (see [2B], Chapter 23). But in the model we are considering here there is by contrast no mystery about it.

To study such 'ontological models' in depth is certainly interesting. But if that is so it seems to me it is essentially for the reason given above, the extremely indirect and disconcerting nature of which must be acknowledged. Indeed, if reality is veiled, it is no longer a matter of discovering *the* truth. Rather, it is solely a matter of verifying, albeit by means of a 'naive' model, the intrinsic non-absurdity of a conception according to which ultimate, non-separable, reality escapes all efforts at descriptions founded on the usual basic notions, but nevertheless gives rise to intersubjective agreement. Undoubtedly this relativises – though without cancelling – the importance of the distinction between the possible kinds of dependence at a distance I have described here. Secondly, it also justifies the fact that in the foregoing discussion I have had recourse, for want of better, to a crude model, whose naivety consists in that, in it, unanalysed *given* consciousnesses *passively* register this or that aspect of the real.

APPENDIX 3

Reverse causality: should it be called causality or not?

In the main text I deliberately avoided mention of ways of conceiving things which I judged, or judge still, to be in need of a definitive coherent formulation. In my view, that was the case until recently with the approach of O. Costa de Beauregard; the notion of causality central to this approach had never been properly defined. This is no longer so, for Costa de Beauregard's recent writings show that he has now succeeded in rendering his conception of causality clear and explicit. However, to do so his way of accounting for phenomena such as the correlations at a distance of the kind associated with Einstein, Podolsky and Rosen (EPR) has had to be elaborated considerably. Let us now see, in their broad outlines, what these clarifications and elaborations involve.

1 The concept of causality

Nowadays Costa de Beauregard suggests [59] that causality should be identified purely and simply with a *conditional probability* (or with a notion of transitional probability that he relates to it). Conditional probability is defined, as is familiar, as the probability that if a particular phenomenon A occurs, then another phenomenon B will 'occur' or 'be observed'. Such a definition clearly does not invoke the temporal order of A and B. Identifying causality with conditional probability therefore makes it possible to give a meaning to assertions that employ the concept of causality; and this regardless of what the temporal order of events may be. In that respect, this identification must be seen as a response – and an acceptable response – by Costa de Beauregard to those physicists who asked what meaning he attributed to the notion of causality when he spoke of reverse causality.

2 The distinction between cause and effect

We must recognise that *de jure* this distinction disappears in this new way of conceiving things. Consider the case in which the conditional probability of A *given* B is 1, and in which the same goes for the conditional probability of B given A (in the language we would normally use, we would then say that between A and B there is a *strict causal relationship*). It is clear that the above definition of causality provides no basis for any possible distinction between A and B that would enable us to say for example that A is the cause of B, and not the other way round. If the attribution of such names is to be meaningful, there needs to be a further stipulation. But it seems that nowadays it is one of Costa de Beauregard's leading ideas that no such stipulation should be provided, or at least that it should not be incorporated in fundamental physics.

3 We can no longer say 'the dice are cast at A (or B)'

Previously, Costa de Beauregard used to make free use of this expression, particularly when explaining correlations of the EPR type. His thesis then was that in the relevant experiments 'the dice were cast' not at the source of the pair of particles but rather at one of the two instant-points A and B where the measurements took place, and that, by reverse causality, the result of the measurement obtained there influenced the source, which in its turn, but this time by ordinary causality, determined the result of the measurement carried out at the other instant-point. Now obviously the claim that the dice are cast not at the source but, say, at A (or B) can be meaningful only if, implicitly or explicitly, we suppose that there is an intrinsic distinction between cause-events and effect-events. In the refined version of his ideas that Costa de Beauregard now puts forward we must therefore renounce such a way of expressing ourselves (and accordingly such a way of representing things). In passing we may note that this enables us to dismiss one erroneous interpretation of his views, in which they are presented as being compatible with a physical realism which ultimately is extremely 'materialistic'. This compatibility is one of the claims made by Ortoli and Pharabod for example, in their otherwise fine book *Le cantique des quantiques* [60], and it must be admitted that Costa de Beauregard's conception in its initial form at least seemed to leave itself open to such a 're-interpretation', albeit that he himself never sanctioned it. The present version eliminates the very possibility of a misunderstanding of that kind. This is all the clearer, given that he himself rejects any idea of the reality of 'intermediate states', which as he insists include in the case of EPR those of the source.

4 'Causality' or mere 'lawfulness'?

I shall conclude by formulating this question. As early as the beginning of the nineteenth century eminent scientists had recommended that the concept of causality be abandoned in favour of that of 'physical law', one of their main arguments being founded quite rightly on the view that the distinction between past and future is *de facto* rather than *de jure*. Once the realistic imagery of Costa de Beauregard's original conception ('the dice are cast' and so on) is abandoned, what remains – and it is the essential part

– is a conception of ultimate reality which is neither embedded in nor even partly constituted by space-time [61], and is therefore, in my language, describable as 'distant' (for my part, I would say 'veiled', but I cannot say whether he would find the term acceptable). In view of this and of the foregoing discussion, it is not at all obvious whether his approach to the general notion of causality and all that goes with it ultimately goes much beyond the classical idea that causality can be reduced to laws.

Costa de Beauregard may in certain quarters be reproached, as I have been, for embarking on the slippery slope of metaphysics. Nowadays, however, to make an honest attempt to construct a realism devoid of all metaphysics seems to be, as we have found, rather akin to trying to square the circle. But that is not to say that none of the attempts at such an enterprise is worthy of our attention. Some of these are fine in themselves, independently of what is always the rather subjective appraisal made of their 'success'. One such, in my opinion, is the all too little known study by F. Bonsack [62], which is well worth noting.

ADDENDUM

*Empirical reality, empirical causality and the measurement problem**

Introduction

A 'realistically minded' physicist of the last century could interpret physics as a faithful – though presumably incomplete – description of 'what really is', without encountering any difficulty internal to science (any objection to this standpoint could come only from *a priori* philosophical considerations). Even today some physicists consider their science should hold fast to this ideal. But most of them assign a more modest goal to physics, and to knowledge in general. Science, they say, (and ordinary knowledge as well) is indissolubly linked with human experience. Once and for all it must therefore give up the unattainable goal of describing whatever some thinkers may mean when they speak of 'reality in itself' or 'reality as it really is'. The task of science can only be a description of the *phenomena*, that is, of things, events and so on, as they are organised by human collective experience. The human means of apprehension and the human means of data processing on which this human experience rests cannot be kept out of consideration and science should not try to do so. Although such a conception was a part of Kant's philosophical doctrine it is considerably less detailed and specific than the latter was, so it is not necessary to be a 'Kantian' to subscribe to it. In fact it constitutes the – explicit or implicit – viewpoint of a great many thinkers and physicists of our times, most of whom have scarcely heard of – and at any rate never took any interest in – the Kantian philosophy. For short let us refer to these people as 'phenomenists'.

Obviously this conception has much in common with the one that may be summarised by the sentence 'human knowledge is just the set of all the ever-successful recipes'; so much so indeed that in the main part of this book no clear-cut distinction was made between the two. The idea which forms the basis of this addendum, however, is that more detailed reflection shows that the identity is not complete; more precisely, it shows that the first view cannot *a priori* be said to reduce to the strict operationalism the second one consists of. There may be a way of 'enriching' pure operationalism. In fact there may even be several ones. But up to recent times only one of them was put into practice. It consisted in supplementing opera-

* This Addendum reproduces, with some additions and changes, an article already published under the same title in *Foundations of Physics* (**17**,507 (1987)).

tionalism with ideas that are compatible with it and that are so natural-looking that they are currently – though implicitly – taken for granted both by the man in the street (including us, in ordinary life) *and* by the philosophers. (It seems that what many philosophers mean when they use the term 'world' is a conception more or less implicitly based on the ideas described here.) These ideas are usually regarded as corresponding to 'rather obvious things concerning objects, their properties and our capacity to know them'. Presumably these ideas were tacitly accepted even by Kant, since they seem more or less implied by his use of the words 'object' and 'reality'. The physicists of course – including the phenomenists alluded to above – also have a strong *a priori* tendency to take them for granted. For what follows it is useful to formulate some of them quite explicitly, even if precision, here as in other fields, entails some degree of ponderousness. They read thus:

Idea A

There must exist a sense of each of the words used such that the assertion 'at any time the centre of mass of any macroscopic solid body has the property of being at some definite place (that is, within some small region *R*) and of not being anywhere else (that is, within regions disjoint from R)' is meaningful and true.□

Such a 'property' is obviously variable in the sense that it can change over time. In what follows the word 'property' is used to mean a property which is variable in this sense. As for the expression 'observable prediction' which occurs below, it means a prediction bearing on the result of a measurement that could conceivably be made.

Idea B

A *property* that a system is said to *have* must be operationally *but counterfactually* defined; that is, its definition must refer to the ways of measuring it but it should not refer exclusively to the cases in which a measurement is actually made, nor to the instrument actually used for this purpose. Instead, when we say that a physical system *S* 'has a property *P*' we mean that *if* anybody came with some instrument suitable for measuring this type of property and applied it to *S* he *would* read on his instrument quite a definite result, identical to the one he and others have read on their instruments in similar circumstances and have conventionally associated with the statement '*S* has property *P*'.□

Remark 1

Of course we say that a system has a property *P* only in the cases in which what we know (for example, about the system preparation) makes us say so. This definition of what we *mean* by attributing a property to a system should obviously not be interpreted as implying the (false) assumption that anything we could measure has some definite value.□

Idea C

Let $\{S_1 \ldots S_i \ldots S_N\}$ be an ensemble of physical systems prepared at time t_o in some given way and let $\{\ldots P_n \ldots\}$ be a set of properties that systems of this type may have. If the computation rules R of a theory T that is acknowledged to be valid (in effect, quantum mechanics) are applied to the assumption 'at time $t > t_o$, S_i has properties P_{ni}, $i = 1, \ldots, N$', if the computation thus made predicts that at some time $t' > t$ some definite set of results will be observed on the ensemble of the S_i, and if the observations actually made at time t' disprove this prediction, the assumption under consideration is thereby disproved.□

Idea C is merely a particularisation of the principle of non-contradiction. But in conjunction with idea B it has an important corollary; for if idea B is accepted, the assumption 'at time $t > t_o$, S_i has properties P_{ni}, $i = 1, \ldots, N$', in no way implies that, at time t, an instrument suitable for measuring the P_ns is actually present. Hence this assumption can then be *formulated* even in the case (which in fact is just the one all of us have intuitively in mind when considering such matters) in which the systems S_i do not interact with any measuring instrument at all during the whole time interval (t_o, t'), thus undergoing during this time interval an evolution that simply obeys the rules R of theory T. For the purpose of testing the validity of the assumption in question we can therefore compare the predictions at time t' derived from applying rules R to it, to those derived from applying these rules to what we know about the initial preparation of the system at time t_o. Hence we can state the following corollary to B and C.

Corollary

Under the conditions considered in stating idea C the assumption 'at time $t > t_o$, S_i has the properties P_{ni}, $i = 1, \ldots N$' is false whenever it is the case that some observable predictions at time t' $(> t)$ derived from applying rules R to this very assumption contradict the observable predictions at time t', derived just from applying rules R to what is known about the initial preparation of the ensemble of systems at time t_o.□

Though science cannot keep the human means of apprehension totally out of consideration, nevertheless the *basic assumptions* of a physical theory should preferably not refer to the ingenuity of present or future experimenters. This is all the more so because this ingenuity has no pre-assigned limits. Hence, even if some physical quantities are not measurable *now*, there is no guarantee that they will not become measurable in some more or less distant future, except of course if such limitations of measurement *follow* from a physical theory T we have good factual reasons for considering as being true (as is the case with quantum mechanics and the Heisenberg relations). This remark actually leads to the acceptance of two ideas D and D'.

Idea D

If we assume a theory T to be true (at present this is the case as regards quantum mechanics) we must consider that the set of observables on the measurement of which the above corollary bears includes all those which are such that the idea they 'could be measured in principle' does not violate theory T (even if actually measuring them would imply such in-practice unthinkable processes as, for example, a substantial human-induced entropy increase of the whole of our galaxy).□

Idea D'

A theory T' propounded for replacing quantum mechanics can, to be sure, imply the non-measurability of some physical quantities as a consequence of its axioms, but it should not be *based* on *a priori ad hoc* assumptions about the impossibility of measuring such and such observables.□

Finally let us add to this set of idea another one, idea E used only in Part 2 of this addendum.

Idea E

It is impossible to influence the past. More precisely, given the initial values of a set of dynamical variables of a system we can, within certain limits, decide on the corresponding final values (and act in such a way that the variables do take these final values) but the opposite is not true. Given the final values we never can freely decide on the corresponding initial values.□

Remark 2

Within a strictly deterministic theory the second part of idea E would be trivial (for the impossibility would then also extend to the future), and within a theory that would not allot to free will some special status of its own, it is doubtful that an idea similar to E could be formulated consistently. But this does not prevent idea E from being a very basic element of our normal way of thinking.□

Remark 3

It would be an oversimplification to say that idea E is disproved right at the start, independently of the other ones listed here, by the mere existence of the so-called 'delayed choice experiments'. [63] The experiments to which this name was given are truly 'delayed choices' only if some additional assumptions are made explicitly or implicitly, such as that of locality (for example, no delayed choice exists in a conception in which the wave function is the sole basic reality and may collapse at a distance).□

Remark 4

There are of course quite a few other 'common sense ideas' we would like to keep. However only ideas A to E are considered here.□

But are all these ideas actually true? More precisely, can we reconcile these ideas – which are precise formulations of a few things we think we know for sure – with the known data and with the predictive rules of quantum mechanics (not to be confused with the interpretation(s) of that theory)? Of course it is a universally accepted fact that a number of basic concepts of quantum physics run counter to our naive intuition. However, this is not what is at stake here. Ideas A to E are very general. They imply no 'naive' postulate bearing on such things as, for example, the particle-like or wave-like nature of matter. And it is neither an obvious fact nor a 'well-known theoretical result' that the ideas considered here are incompatible with the quantum rules and with the data. Indeed, a number of theories (mainly measurement theories) were recently propounded that claim to reconcile the data and the quantum rules with natural-looking general ideas about macroscopic objects and about the meaning of such general notions as that of a property; it might therefore *a priori* have been expected that these theories would, in particular, be compatible with ideas A to E. In Part I of this addendum however it is shown that this is not the case, at least as regards the theories of which I am aware.

Part II of this addendum can be read independently of the first, if one simply accepts the negative conclusion just stated. Its subject matter lies on the borderline between philosophy and physics since it is directed to the question of ascertaining whether or not the conclusion in question (assuming it is general) entails the necessity of retreating to the philosophical position of strict operationalism (science and knowledge, even of ordinary things, viewed as mere prediction recipes). It is not *a priori* obvious that this is the case, for it is not inconceivable that modifications can be brought to the set of ideas A to E which will not amount to discarding them completely, and it may be hoped that, presumably at the price of changing some of our ingrained views about time, space and macroscopic objects, the new set of ideas thus obtained will still, to some extent, play the same role as the old one, namely that of making factual knowledge something somehow more basic than just a set of good recipes for prediction. Accordingly, Part II addresses the question whether or not a new conception of phenomena, or of empirical reality, can emerge, one that will play more or less the same role as the one entertained by the phenomenists of the past, but without contradicting the data and the quantum rules. In particular it shows that for this purpose the notion of an 'empirical causality' endowed with somewhat surprising features needs to be introduced.

To be sure, an appreciable number of physicist readers will consider this second part much too 'philosophical' for their taste. But not all physicists shun philosophical problems. Many are aware of the fact that contemporary physics unavoidably raises some important problems of such a nature.

Part I

Our purpose in this part is to examine a number of recent theories that could *a priori* be considered to reconcile quantum physics with what we are tempted to regard as 'most obvious things concerning objects and their properties', and to show that this reconciliation fails by proving that the theories in question are incompatible with ideas *A* to *E*. This, in a way, is a continuation of a former work (see [2B]) in which I did the same as regards more ancient measurement models. The content of this section should not be considered as a criticism of the theories under examination, since the authors of most of them more or less explicitly say that they discard some of the ideas in question, in particular idea *D*. Rather, it should be viewed as an attempt at a better determination of what kind of picture of empirical reality we are allowed to retain when we take these theories into consideration.

One of the main problems quantum theory has to deal with is how to account for the measurement process. And one of the most basic difficulties (some even say *the* basic difficulty) of quantum measurement theory concerns the fact that if we consider a pure state ensemble of quantum systems, each of which interacts with an instrument appropriate for measuring a certain physical quantity, the instrument is expected to register some definite result in each case and afterwards to be in a state corresponding to this result. In a case in which the initial ensemble of measured systems is in a superposition of several distinct eigenstates of the quantity in question, this implies that the final ensemble of the combined systems (measured system plus instrument) must be a mixture, even though the initial ensemble of the measured systems was a pure case. Hence some cross terms must somehow disappear in the representative statistical operator ('reduction of the wave packet'). At any rate, this appears as a necessary condition any measurement theory must fulfil and it is the problem of how actually to fulfil it that consitutes the subject matter of so-called quantum measurement theories (whether it is also a sufficient condition is not obvious, for a measurement involves just *one* instrument interacting with *one* system, not an ensemble of instruments *plus* systems; but we shall barely touch on this point).

1 Machida and Namiki

The purpose of the Machida and Namiki [64], [65] theory is to show that the troublesome cross-terms vanish due to the macroscopic nature of the measurement apparatus. More precisely, it expresses this nature by describing the apparatus in question by means of a continuous direct sum of many Hilbert spaces. Here we shall not need any very detailcd description of the sequence of formulae through which this theory is made explicit; the reader is referred to the original articles for systematic information. We are only interested in trying to make as precise as possible the nature of the assumption on which this theory rests.

To prove that the cross terms under consideration do vanish the authors resorted to the well-known idea of averaging over the phases; however they used this idea in an original manner, which makes their argumentation worth much closer attention than the earlier attempts by other authors. Briefly summarised their idea is that although the relevant part A of the apparatus may be quite small, nevertheless it is macroscopic; that we therefore cannot sharply determine its energy and particle number, even by using the longest time interval available to us; and that consequently this part should be represented by a statistical operator having the form

$$\rho_A = \sum_{N \in (N_0, \Delta N)} W_N \rho_N^A \tag{1}$$

where $(N_o; \Delta N)$ stands for an interval with width ΔN around N_o, W_N for a positive weight factor ($\sum_N W_N = 1$) and N for the particle number of A.

In a second step they take advantage of the fact that N_o is very large while $\Delta N/N_o$ is very small and they replace the discrete sum occurring in eq. (1) by

$$\sigma^A = \lim_{N_0 \to \infty} \rho^A = \int dl W(l)\, \rho^A(l) \tag{2}$$

where $l = aN$ and where $W(l)$ is a normalised weight function centered on $L = aN_o$ with a width $\Delta L \simeq a\Delta N$. When the As thus described are made to interact with the microscopic systems, initially assumed to be in a pure state described by $\psi = c_1 u_1 + c_2 u_2$ (the case of a spin ½ system with σ_z eigenvectors u_1, u_2 is considered here for simplicity, σ_z being the observable to be measured), the density matrix of the (ensemble of the) combined system(s) becomes in the authors' model (the details of which we skip here):

$$\Xi_t = \sum_{i,j} c_i c_j^* \, \Xi_t^{i,j}, \qquad i, j = 1, 2 \tag{3}$$

where the integrals

$$\left[\int dl_i \, W(l_i) \, e^{-ip_i l_i} \ldots\right]\left[\int dl_j \, W(l_j) \, e^{ip_j l_j} \ldots\right] \tag{4}$$

appear in the terms $\Xi_t^{i,j}$ for which i≠j (the dots representing smooth functions of the l's) whereas *only* integrals bearing on smooth functions of the l's appear in $\Xi_t^{i,i}$ (i = 1, 2). In eq. (4) the p_i play the role of effective momenta ($\hbar$ = 1) and are determined by the details of the model. In the limit in which $p_j \rightarrow \infty$, the integrals in eq. (4) vanish (Rieman–Lebesgue theorem) so that the desired result is achieved.

As stressed by the authors themselves, the averaging procedure described by eq. (2) is quite essential for their theory to go through. As we saw, its justification is based on the fact that in practice *we* cannot determine precisely the number N of particles composing A, and on the view that the large but finite mean value of this number can be replaced by the limit $N_O \rightarrow \infty$. Clearly, a reference to the limitation of the human abilities is involved there. This reference contradicts idea D. Moreover, even if the quantities p_i are large they are not infinite so strictly speaking it cannot be asserted that the values of the integrals in (4) are exactly equal to zero. For these and similar reasons it was stressed by H. Araki that the reduction of the wave packets taking place in this theory is only approximate.

Since the reduction of the wave packet is only approximate there is no reason to expect that for *all* possible observables the predictions derived from the reduced density matrix are to be the same as those correctly derived from the non-reduced one. Admittedly the theory shows that this must be the case for all the observables that we, as human beings, can reasonably expect to be able to measure in practice, and this may be viewed as an essential achievement. Very justifiably the authors demand no more. But in fact there must exist Hermitian operators for which these two types of predictive recipes lead to different results. There is no reason of principle for assuming that these hermitian operators do not correspond to observable quantities. Under these conditions it must be concluded that the reviewed theory conflicts with the set of ideas A to D. More precisely, in a case such as that of a Stern–Gerlach measurement for example, if by assumption idea D is true there is no state reduction. In other words, after the spin–instrument interaction has taken place, a pointer geared to register which one of the counters has fired cannot lie in any one definite graduation interval, contrary to idea A (if it did then, on an ensemble, some of the physical quantities that are measurable in principle would not have the values quantum mechanics enables us to derive from the known initial conditions; see ideas B and C and the corollary).

2 Araki

In non-relativistic physics the number of particles composing a system cannot change and there exists no observable quantity that would allow us to distinguish a 'pure state' represented by a linear superposition, with coefficients c_N, of state vectors describing states with different particle numbers N, from a 'mixture' with weights $|c_N|^2$ of states with definite particle numbers. In such cases it is convenient (see above) to consider several Hilbert spaces H_N, indexed by N, the observables being described by self-adjoint operators defined within each H_N (that is, not connecting different H_Ns) and it is said that a superselection rule exists. When N is unknown (with mean value N_O) the description of a state of the system by a statistical operator then takes the form of eq. (1). Taking the limit $N_O \rightarrow \infty$, $\Delta N \rightarrow 0$ as in the Machida–Namiki theory then obviously corresponds to going over from a discontinuous to a continuous superselection rule. It is therefore clear from the content of foregoing pages that the consideration of a continuous superselection rule is an essential component of the Machida–Namiki theory.

On the other hand, the introduction of a continuous superselection rule in a measurement theory is not sufficient, as we saw, to make the latter an 'exact' theory, where 'exact theory' means here a theory compatible with ideas A to D. It is on this question (of how to make an exact measurement theory) that Araki recently made an important contribution [66]. Again, we refer to the original work for details and merely summarise the results. There are three distinct ones:

(i) A continuous superselection rule can give rise to reduction of wave packets in a quantum mechanical *separation* procedure (in contrast to a measuring procedure) in the infinite time limit (in Araki's terminology a separation procedure differs from a measurement procedure in that the instrument is not brought in distinct states by it).

(ii) The reduction of wave packets is impossible in the case of a *discrete* selection rule (or of no selection rule whatsoever).

(iii) Even in the case in which a continuous superselection rule is present, the reduction of the wave packets and the measurement procedure proper must proceed in two distinct steps. No set-up can exist that would produce them simultaneously. *First* the reduction should take place (and it takes an infinite time), and *next* the measurement should be made, using some conventional apparatus. At least, in the author's words, this seems to be the best that can be achieved. For completeness let it be noted that the author does not consider the cases in which the measurement alters the state of the measured system (second kind measurements). However, it is doubtful that taking such cases into consideration could alter the author's general conclusions.

As regards the continous superselection rule, it is of course not surprising that it should be associated with a classical variable. This variable is N in the Machida–Namiki case. In Araki's model it is a classical magnetic field **h**. The general argument concerning this association is, as we know, that the variable that serves as a label for the various Hilbert spaces obviously has a sharp value on *any* pure state. By definition such a variable is called 'classical'. Hence a classical variable is necessarily associated with a superselection rule. Now, since the presence of a superselection rule is a necessary condition for an exact measurement model to be possible, it follows that such a model is possible only within a general theory that enlarges elementary quantum theory by incorporating classical variables. In principle this is not a difficulty since the theories based on the algebra of observables do precisely this (some comments on this point appear below). On the other hand, an essential feature of Araki's model is that the possible (classical) states of the magnetic field system should be described by *continuous* functions (thus barring δ functions). This rules out the possibility for this state to be a 'pure state', since a pure classical state (in Segal's sense) corresponds to a δ function probability distribution. In other words the possibility that the magnetic field **h** has a sharp value is definitely excluded. As the author puts it, 'this restriction may be interpreted to represent our inability to control the external magnetic field with absolute accuracy'.

With respect to the question here investigated Araki's article is important, to begin with, in that it proves that the ordinary quantum measurement theories (including the Machida–Namiki one as we said) are incompatible with ideas A to D. According to it, in our search for models that would be compatible with this set of ideas all we are left with are candidates of a rather unusual type, in which a reduction procedure is made *before* the actual measurement takes place. But when all is said and done, even these models cannot be reconciled with the set of ideas in question. The reason for this can be formulated in two stages, which taken together constitute a strong argument. The most obvious one is that the reduction procedure takes an infinite time and we have only a finite time at our disposal. Trivial as it may appear, this remark is nevertheless most significant. The only way in which we could perhaps hope to be entitled to overlook it would be to argue that somehow and for some reason we are allowed to replace large numbers (here, times that are long compared with average atomic times) by infinite ones. But if we indulge in this mathematically questionable manipulation it is difficult to understand why we should refrain from replacing very small numbers by zero, and in particular why we should exclude the 'pure states' of the **h** system. The advocates of mathematical rigour cannot be one-sided. If they forbid the second move they must forbid the first one too.

3 The environment theories

Briefly summarised these theories consider the real cause of the 'wave packet reduction' to be the fact that the instrument is not strictly isolated from its environment. Some of them (see for example [67] and [68]) moreover claim that the environment also ultimately determines *what* physical quantity is actually measured, by creating a 'preferred basis'.

Here we are only interested in the reduction problem, and again we need not enter into the finer details of these theories. In them the mechanism of wave packet reduction is as follows. To the right-hand side of an equation such as (1) above they add some term describing the environment variables. To account for the fact that we do not measure any of these variables they then prescribe taking the partial trace of the full statistical operator over the environment Hilbert space. It can then be verified that when the time t becomes large the coefficients of the cross terms $c_i^* c_j (i \neq j)$ all become extremely small.

Our purpose here is not to appraise the value of the environment theories as compared with other quantum measurement theories. It is just to investigate whether these theories are compatible with ideas A to D. It turns out that they are not, and, here again, it must be stressed that their authors do not claim they are. In Zürek's words 'information is not destroyed, it is merely transferred'. The point is that it is transferred to entities that, for practical reasons, human beings are quite unable to measure. From a strictly operationalistic point of view this is obviously quite sufficient, and we must therefore say that these theories do indeed attain their objective. But this of course does not imply that they are compatible with ideas A to D. The reason why they are in fact not is qualitatively the same as in the case of the Machida–Namiki theory. Within any finite time the coefficients of the cross terms never strictly vanish. Some self-adjoint operators therefore exist, the mean values of which are not the same on the actual ensemble as on the reduced one. If the environment is composed of N atoms and if N is large the difficulty of measuring the corresponding observables must be great, and it may increase beyond limits; that is, it may become a strict impossibility, in the idealised case $N \to \infty$, $t \to \infty$. Admittedly, if we consider as conceivable only the measurements that have a given, finite degree of difficulty we may then be able to find a number N_O such that whenever $N > N_O$ the cross terms are not detectable. But on the other hand, to any given finite N there must correspond a degree of difficulty such that if we may conceive of measurements the difficulty of which exceeds it, the cross terms *are* detectable. The conclusion is then the same as in the previously considered models. If idea D is to be kept, measurements of any degree of difficulty are conceivable, the cross terms *are* in principle detectable, the quantum mechanical predictions concerning the observables which makes it possible to detect these terms are incompatible with the assumption that idea A is valid, and hence idea A cannot be kept.

A general remark

Because of its relevance to the problems investigated here I shall restate here a remark already formulated elsewhere (ref. 2B, p. 195). It bears on the smallness of the difference between the observable predictions derived from the non-reduced statistical operator, and from the reduced one. In view of this smallness we could be tempted to say that after all the state of affairs considered here does not significantly differ from the one existing in classical physics since, there also, objects can never be considered as being 'strictly separated and non-interacting'. Could the fact that if idea D is kept, idea A cannot strictly be maintained (and that therefore a pointer cannot properly be said to *be* in any one graduation interval) be just another instance of this trivial impossibility? The answer is *no*. The two cases are in fact quite different. In classical physics an approximate statement S about a system T can always be replaced by a strictly true statement S'. For example, if T is composed of two distant objects, the approximate statement 'the two objects have practically no interaction with one another' can be replaced by a precise statement of the form 'the interaction between these objects is smaller than x'. On the contrary, concerning the combined system (including apparatus and environment) considered at a time $t < \infty$ after the interaction, there is no precise statement embedding idea A that can be reconciled with idea D. The precise statement that such and such a quantity is smaller than a certain value bears only on those mathematical artifacts that ensembles amount to. They do not bear on the actual combined system.

4 The algebraic theories

Some authors have claimed that a number of conceptual difficulties appearing in 'ordinary' quantum mechanics could be removed by resorting to the more elaborate theory known as algebraic quantum mechanics. Araki's theory may in some sense be considered as belonging to this realm. Hepp's theory [69] is the best-known example of a measurement theory constructed along such lines. However, this theory was discussed by Bell [70] and the sequence of these two articles makes it clear that, in fact, the reservations formulated above in connection with the environment theories also apply to Hepp's theory (a more detailed study of this point was already presented in ref. 2B). As a rule the algebraic theories bring in new interesting possibilities essentially when the involved systems have an infinite number of degrees of freedom; and in the case of the algebraic measurement theories one should even say: have an infinite number of *relevant* degrees of freedom (such as the 'number of atoms' considered in the foregoing paragraph). But while the existence of systems with an infinite number of degrees of freedom is an undisputable fact, it is, in most theories, an open question whether the systems endowed with the role of instruments have, under sufficiently realistic experimental conditions, an infinite number of *relevant* ones. Unfortunately, the highly abstract nature of the language in which the algebraic theories are couched makes this point still more obscure. It is therefore not surprising that when concrete examples of algebraic measurement theories are sought, models emerge in which the infinity in question appears as an idealised limit, thus paving the way to the above mentioned criticisms.

This remark give a glimpse at the difficulty of constructing satisfactory quantum measurement theories, even using the algebraic approach. Nevertheless Lockhart and Misra [71] recently put forward an interesting algebraic quantum measurement theory aimed at refining Hepp's theory in such a way that Bell's criticisms should no longer apply. Starting from conventional quantum mechanics these authors constructed a formalism in which, when the dynamics of the (macroscopic) measurement instrument (considered as an infinite system) has enough complexity and instability the quantities pertaining to the instrument (and instrument plus system) which are measurable 'in principle' can be, and are, assumed to constitute a time dependent strict subset Σ of the 'observables' of conventional quantum mechanics. In this formalism the system *plus* instrument 'states' (in Segal's sense) must then be redefined on this subset. The time evolution law of these states differs from the quantum mechanical one, and is irreversible; and, when the instrument dynamics is complex enough, the states in question spontaneously evolve from a pure case to a mixture, in conformity with the basic necessary condition appearing in the measurement problem.

Also in this case, the readers interested in the quantitative developments of this very advanced theory must be referred to the original article. The foregoing description is but a short qualitative account of its general structure. However, this description is sufficient for discussing the question of the compatibility of this theory with the set of ideas A to D′ (we leave aside, because of its complexity, the question concerning the infinity of the *relevant* degrees of freedom).

At first sight the question would seem to be settled by the negative. For we see that the assumption according to which some 'observables' cannot be measured is quite basic to the theory, in the sense that it is not a mere consequence of other ideas taken as axioms but constitutes *per se* the main idea on which the whole theory is constructed. This is already an indication that the theory in question conflicts with idea D′ (idea D′ is here the relevant one since what is actually proposed is to change quantum mechanics). The point, however, is worth examining further, for we are told that the observables not in Σ cannot be measured 'in principle'. As the authors write, 'what cannot be measured depends to a large extent upon the Hamiltonian dynamics of the system, and not upon the present state of the art or the ingenuity of the experimenters'. It must be acknowledged that this is indeed the assumption made. On the other hand, for the purpose of comparing this theory with idea D′ the meaning of the expression 'in principle' must be clarified, and for this the restriction 'to a large extent' appearing in the quoted assertion is significant. In fact, neither in the theory itself nor in the example that illustrates it in the quoted article, is the precise nature of the observables about which non-measurability is postulated derived from anything known. These observables are just chosen in an *ad hoc* way. Hence the expression 'in principle' cannot refer to any basic axiom of the theory and idea D′ is indeed violated. (Note, in addition, that even if the choice of the observables in question were determined by the dynamics, we would still have to do with a theory one basic assumption of which depends on the dynamics of the systems involved, and which therefore could not be axiomatised as 'framework theories' must be, see Chapter 11, Question 19). What the theory seems actually to provide us with is but a general formalism within which some assumptions about the non-measurability of such and such observables can consistently be formulated and have the desired effects. Let it be observed, moreover, that since the non-measurability in question can in no way be considered as a consequence of the general physical theory, it seems natural to interpret it as due to limitations in the aptitudes of mankind.

Conclusion

There are three conclusions to this Part 1. The first one is, of course, that the set of ideas A to D′ turns out not to be compatible with the idea that the quantum mechanics is the correct theory and with the recent measurement theories reviewed here (just as a basically equivalent set of ideas turned out to be incompatible with the older measurement theories, examined in my earlier book (ref. 2B)). Of course we could not study *all* the measurement theories, existing or as yet to come, but it can nevertheless be concluded that the plausibility of the assumption that a new measurement theory, compatible with the set in question and with quantum mechanics as it stands, will eventually turn up is becoming extremely small. We take no great risks in discarding this assumption.

Should we then consider replacing quantum mechanics by some more refined theory that would account for the irreversibility of measurement by making irreversibility a basic feature of reality? The second conclusion deals with this proposal. In the Introduction we pointed out that, in order to meet the requirements of the phenomenists who are dissatisfied with *pure* operationalism, any new basic theory should be compatible with idea

D′. However the one that has been proposed along the lines just sketched, though interesting and possibly true, does not comply with this request. Whether future ones will remains doubtful.

The third conclusion is that, as previously noted, the negative results just mentioned cannot *a priori* be considered as compelling us to fall back on a strictly operationalistic conception of knowledge (science as a mere set of recipes); for it may conceivably be that the set of ideas A to E can be replaced by another one that could be reconciled with quantum physics. The quest for this new set is the theme of the second part.

Part 2

For partly philosophical and partly physical reasons unfolded in the main text, I consider that the notion of an independent, veiled reality is necessary. However, this reality is *not* the subject matter of this Addendum. As already noted our problem here is to steer a mid-course between the conception that the purpose of science is to *unveil* this ultimate reality (the view we called 'physical realism'), and the one according to which science and knowledge are but successful *recipes* for predicting what a human community will, in given circumstances, eventually observe (a view sometimes called *strict* or *philosophical operationalism* or *instrumentalism*). At this stage many physicists will insist that strictly speaking the problem in question is not a scientific one. This may well be admitted (defining science lies beyond the scope of this book) but at the same time it must be stressed once again that the philosophers are not adequately equipped for studying the question unless they have learnt physics. Hence this study *calls* for physics. Such an observation may be considered as an adequate justification for the fact that at least *some* physicists do take some interest in this question.

The mid-course considered here consists in trying to enrich the purely operationalistic viewpoint by adding to it some ideas, much in the same way as the 'phenomenist' physicists of the past instinctively added, as we saw, some ideas to this same operationalism, and in this way wrought a conception of what could be called an *empirical reality* somehow endowed with appreciably more substance than a set of recipes can have. The difference is that most of the phenomenists in question did this implicitly, guided as they were by 'common sense', whereas we must proceed in full awareness of every step we take (another difference is of course that, as we also saw, the older phenomenist viewpoint was still too near to naive realism to accommodate quantum physics and that our objective is to build up a new conception that should be compatible with it).

Notwithstanding what was just mentioned, it is clear that our new, constructed, notion of an *empirical reality* can be of any significance only if it keeps as close as possible to our 'apparently obviously true' ideas, concerning, in particular the macroscopic world. This implies among other things that we should try to keep as much as we can of the substance of ideas A and E (see the Introduction). As regards idea B, the only possible alternative to it seems to be to replace counterfactual definitions by 'partial definitions' in the sense defined in Chapter 3. But to apply such definitions to macrosystems would mean that the use of the conditional mood should also be strictly avoided in connection with the properties of these systems, and this seems awkward. Moreover, since we have no fully-consistent criterion for distinguishing macro from microsystems the rule according to which the partial definition procedure should only be applied to the properties of the microsystems would be an ambiguous one, if not in practice then at least in principle. The considered alternative is therefore unattractive.

Hence let us keep idea B. Since idea C is just a particularisation of the principle of non-contradiction it must also be kept. Admittedly, we could consider solving the measurement problem by replacing quantum theory by some new, basically irreversible theory. This however seems to be a heavy price to pay, especially since the only theory of that kind presently at our disposal still violates one of the ideas that it would then seem quite natural to keep, namely idea D′. Hence finally it seems that the best remaining possibility is to modify idea D. Quantum mechanics can then remain unaltered.

The axiom of empirical reality

The substance of idea D can be expressed by asserting that when discussing questions of principle no limitation should be set on the sophistication of the measurements whose possibility is being considered, as long as no quantum mechanical principle is violated thereby. It was pointed out in Part I that the notion of an unbounded degree of sophistication is obviously an abstraction, but that this abstraction is a natural counterpart to the one that consists in considering as infinite some measurement times that are actually finite, or some numbers of particles that are finite too. The above proof of the fact that the measurement theories studied there are incompatible with ideas A and D taken together rested on the observation that, for any finite time – long as it may be – or for any finite number of particles – however large – observables and measurements can in principle be conceived, the sophistication of which exceeds the limit up to which measurements detect no difference between the reduced and unreduced statistical operator; so idea A is falsified as regards such things as instruments pointers – at least in some cases – and cannot therefore be considered as being universally true. To avoid being forced to this conclusion it is necessary to find a substitute for idea D. And the first point we want to make here is that this can be done in quite a precise and unambiguous manner, just by stating explicitly an idea which in fact lies at the basis of most quantum measurement theories but which is always kept implicit. This idea is best expressed as an axiom, which may be called the axiom of empirical reality. It reads thus:

> *Axiom of empirical reality*
>
> A theory of empirical reality is obtained by postulating (a) *that replacing very large times by infinite times and/or very large particle numbers by infinite numbers is a valid abstraction; and* (b) *that on the other hand the possibility of measuring observables exceeding a certain degree of complexity is to be considered as non-existent, even in matters of principle, even though this non-measurability does not follow from the theory and even though the only way we have for making it compatible with quantum mechanics seems to be to ascribe it to some basic inaptitude of men.*

Clearly, when in the set of ideas A to E idea D is replaced by the *axiom of empirical reality* the incompatibility between the set in question and the predictive rules of quantum mechanics vanishes. This removes the main difficulty that prevented us from saying, in accordance with idea A, that within the system-plus-instrument(s) ensemble, and when the measurement is over, the pointer of any instrument has some definite macroscopic location. It is to be observed that the axiom in question is in fact an explicit formulation, not only, as we said, of the implicit ideas lying at the root of various quantum measurement theories but also of many of the arguments used in physics and consisting in 'approximating' large numbers by infinity. Reference is made here, in particular, to the approximations that make it possible to 'bridge the gap' between quantum physics and chemistry (see for example [72]) by making use of such methods as the Born–Oppenheimer procedure and, more generally, to the 'approximation' that makes it possible to connect quantum and macroscopic physics by somewhat artificially going over to systems with an infinite number of *relevant* degrees of freedom, thus endowing the algebras of observables with a 'centre' (see again [72]).

Remark

The axiom of empirical reality removes a difficulty of measurement theory. But to be fully satisfactory such a theory should also describe the mechanism by means of which an *individual* system eventually 'picks out' at random a *definite* final state (here 'system' means of course a composite system 'measured system plus apparatus'). It seems fair to say that most theories do not even touch on this problem and that none has mastered it (this is a further argument for not identifying empirical reality with independent reality, adding to all those already mentioned; see Chapters 6, 8 and 10).

On the status of the concept of empirical reality

We asserted above that the axiom of empirical reality makes it possible to attribute a definite macroscopic position to any macroscopic object, including the 'pointer' of an instrument; and that this even covers the cases in which the instrument previously interacted with a quantum system not initially lying in one of the eigenstates of the quantity the instrument can measure. But let us try to be precise as regards the meaning of this assertion. It is counterfactually defined – as idea B requires it to be for any assertion bearing (as this one does) on a property of a system – but it must be granted nevertheless that the wishes lying at the root of our intuitive reasons for demanding counterfactuality are not fully met by the assertion in question. In fact, these wishes and reasons revolve around the idea that even if we cannot know 'the ultimate nature of things', still the macroscopic objects can be said to be fully mind-independent and can be described as such. With the *axiom of empirical reality* replacing idea D this view can no longer be strictly maintained since, at a crucial point in the whole picture, this axiom limits the sophistication of the measurements that are taken into account, by referring to practical limitations in the abilities of the human species.

The status of the concept of empirical reality therefore turns out to be a subtle and hybrid one. As it is described here, this notion clearly has much in common, if not with our naive idea about 'real things', at least with the view the most thoughtful minds have taken of what should be called real. As recalled above, it makes it possible to endow the algebras of observables with a centre, in such a way that, in many important cases, we can use in an unambiguous way expressions of the type 'at such and such a time such and such a system *has* such and such properties'; and for this reason it goes much beyond a mere set of recipes, in the direction of realism. On the other hand, as we just saw, it definitely does not go as far in this direction as – *a priori* – the upholders of philosophical realism, and even most phenomenists, would have wished we could go. In fact it allows us to use the verbs 'to have' (in assertions such as 'this system has this property') and 'to be' (in assertions such as 'this system *is* in such and such a domain of space') only in somewhat weakened senses, since in some cases some measurements that are not ruled out by any theorem, that we ruled out just 'by decree', would falsify what we then say.[1] Empirical reality may then be defined as the set of all the subjects of the verb 'to be', taken in this weakened sense, whereas independent reality, (a notion on which, as previously noted, this Addendum touches but subsidiarily) would be the set of the subjects of the verb 'to be', taken in the strong sense in which the realists of yore took it (supposing of course reliable sentences could be constructed with such subjects, account being taken of what we know now).[2]

Empirical causality

As already stressed, the concept of empirical reality can constitute a significant complement to pure operationalism only to the extent that it makes it possible to refine the ideas we naively consider to be obvious, while keeping much of their spirit. In view of this we still have an important point to explore, for there is one idea that *a priori* we would very much like to keep. This is our usual way of *explaining* what takes place. It is a fact that we account for most of the details we observe (either in daily life or in macroscopic investigations of a scientific nature) by attributing them to the occurrence of definite events belonging to the past history of the presently-observed system or systems, hence in particular to events on which our *present* observing procedure has no effect (idea E). Suppose for example that a counter placed within one of the outgoing beams in a Stern–Gerlach device is connected to a bulb so as to make the bulb light up some time (say a few seconds) after it has fired. In such circumstances we like to *explain* our observation at some given time of the bulb lighting up by saying that at a somewhat earlier time the counter *did* actually fire, and induced the bulb to light up, just as we explain the presently observed U-shaped valleys by saying that during the ice ages glaciers were *actually* there and did give them their present shape. One of the psychological reasons why at first sight we consider as grotesque the idea of a counter being in a quantum superposition of states (states of having and of not having fired) is just that this idea *prevents* us from mentally conceiving of a chain of relationships between observed macroscopic effects and some (one or several) *definite* macroscopic causes, thus throwing doubts at quite a deep level on an essential element of the mental scheme by means of which we explain to ourselves (and to other people) those things that we believe we understand.

But are we really prevented from building the chain in question? At first sight the adoption of the axiom of empirical reality would seem to remove the difficulty at one stroke. For if, in the example considered, we apply this axiom to the bulb lighting up (thus making this lightening a definite event, which does or which does not take place), we of course also apply it to the firing of the counter, which is also a macroscopic event, and which thereby is made definite too. However, this argument does not completely clear up the matter at hand, as can most easily be seen by considering a hypothetical case in which, instead of having happened just a few seconds before the bulb lighted up, the firing of the counter took place, say, a million years ago, and in which the 'causal chain' between these two events involved many intermediate steps. The point is that applying the axiom implies shaping the empirical reality according, in part, to human decisions (about the level of complexity at which experiment should stop) or at any rate according to what the human abilities are. In the circumstances just described, applying this axiom to the counter means therefore that the empirical reality of the past – that empirical reality we refer to when we speak of past macroscopic events such as the formation of the Sun and so on – can in some way and in some cases depend on some decisions (in the above sense) or abilities of the human beings.

Macroscopic causality understood in this weakened sense may be called *empirical causality*. This conception is a (subtle) alteration of idea E since in it, as we just observed, the individual events composing the empirical reality of the past, such as, in the above example, a counter firing or not firing, in a way depend on us. Not, of course, in the sense that we could *decide* whether they occurred or not but just through the fact that, at the considered time and in the strong sense of the verb 'to be', the counter actually 'was' neither discharged nor undischarged.

A priori this seems to raise the following puzzling question: why is it the case that (assuming we have enough information) we all, on the basis of what we see, *agree* to say either that the counter fired or that it did not? Basically, however, this question is neither more nor less puzzling than the one concerning the agreement between several persons who *now* observe a counter used in such a Stern–Gerlach experiment. The fact that we do agree with one another as to what we see in instances such as this one (and more generally in observed events taking place in macroscopic, empirical reality) is not as trivial as long practice induces us to think. Indeed it is an enigma (see the last *remark* above). But this, after all, is not surprising since the explanation formerly given (which seemed obvious) of the agreement in question was based on a specific assumption (which *also* seemed obvious) about the independent reality of the localised objects, since we now know that this conception was too naive, and since we do not know for sure what other conception should take its place (relative state theory, non-local hidden variables theories, non-mathematical descriptions: there are possibilities but certainly no definite knowledge).

As regards empirical causality another riddle is of course present. This is the one concerning the EPR correlations between the results of measurements that are spatially separated. As Bell's theorem shows, there is no way of accounting for these correlations by signals or by energy transfers travelling at subluminal or luminal speed. Yet measurement results (firing of counters and the like) belong essentially to the empirical realm and empirical reality should not be endowed with non-separability features if we want it to remain close to anything we experience (the notions of *local* objects and events are so essential to it, as we saw, that merging them into a 'sea' of non-separability and holism would deprive the concept of empirical reality of its significance). Here again, to search for an answer we must look in the direction of the notion of independent reality. More precisely, we have at our disposal a well-defined set of rules, namely the principles of quantum mechanics, that allow us to predict without any ambiguity what correlations of this type will eventually be observed. Raising some or other of the algorithms (kets and so on) used in formulating these rules to the status of an exact description of independent reality would yield an 'explanation' of these correlations. This however cannot be considered as a final answer since (as we saw in Chapter 10) any postulate according to which such and such quantum algorithms are exact descriptions of independent reality leads to severe difficulties. Hence, here again, it must be acknowledged that we have good *descriptions* but no fully-reliable *explanations* of the facts under consideration.

Empirical versus independent reality and conclusion

A remark on independent reality is here in order. The pervading practical and scientific importance of empirical reality contrasts with the elusive role of the concept of independent reality. Even at present, many physicists who write the word 'reality' actually mean 'empirical reality', and quite sincerely believe they neither need to nor in fact do make reference to what we called 'independent reality'. However, as we remarked in Chapter 10, there exists a simple criterion that makes it possible to check whether or not they actually do, in this respect, what they believe they are doing. It consists in examining whether or not they use the word 'nature' and in what sense. It will then easily be discovered that they all do; that, in the contexts in which they use this apparently familiar word *nature*, it can only mean 'independent reality', and that they could not express their ideas if they abstained from using it. Both the concepts of 'empirical reality' and of 'independent reality' must therefore be retained.

Surely, some who long for a mathematically expressible solution to the basic troubles may dislike the one proposed here on the ground that introducing as we do a mind-dependent empirical reality into the picture is to give too great a role to intersubjectivity. In my opinion this objection is not well founded. The reply is that what human beings actually find does not always coincide with what they would like to find, that mathematically-expressible strict solutions of the 'realistic' variety seem to be escaping us, and that it is rationally possible that the role in question, when its ins and outs are clarified and made, as here, an inherent part of the theory, should in fact constitute *the* correct answer.

To conclude: it seems that the project of steering a mid-course between metaphysical realism and strict operationalism without violating quantum physics is not an impossible one. The central piece of our proposal for carrying it out is the axiom of empirical reality, replacing idea D and making it possible to keep idea A, (localisation of macro-objects), with however the specification that the verbs 'to have' and 'to be' must be taken in their weak sense. Idea E (no influence of the present on the past) then remains tenable, provided that the word 'was' should also be taken in the weak sense, and provided the usual (and reasonable) requirement that intersubjective agreement should not only be described but also *explained* should be watered down. The assertion that the macro-objects are mind-independent can then be retained, even though this is possible only under the (here steadily recurring) condition that the verb 'to be' is given its weak sense. Similarly, our normal way of explaining macroscopic events by sets of definite macroscopic causes can be preserved, be it only in the form of 'empirical causality', without contradicting basic, that is quantum, physics. As regards the internal consistency of man's rational thinking this result is most desirable and, even if the price paid for it here may look quite high (watering down of quite normal requirements, see above), still, the fact that it is obtained may be considered as reassuring.

Notes

1. The positivism of the physicists

1 Briefly speaking, cosmological theories in a way resolve the problem by introducing local frames of reference within which the constituents of the Universe have on average zero or very low velocity. However such theories in no way lead to restoration of the pre-relativistic concepts of universal Euclidean space and absolute time.

2 Some commentators refer to instrumentalism as a kind of device for achieving an economy of thought that would discredit the great syntheses of mathematical physics. That sense of the word clearly has no connection with the one I use here.

3 Occasionally I shall speak of 'instrumentalism in the strict sense of the term' to distinguish it from 'methodological instrumentalism'. In most cases though the context will make this unnecessary.

4 This doubt should not be confused with doubts about the beneficent or harmful nature of scientific research. They too exist, but plainly are quite different from the kind of doubt I refer to here.

5 *In Search of Reality* [2]. By contrast, I use the term 'strong objectivity' to refer to a statement that contains no reference – not even implicit ones – to the community of 'observers'.

2. Positivism and fallibilism

1 Here we touch on the source of one of the greatest problems encountered in teaching quantum physics. Until it comes into contact with quantum physics, the human mind, whether trained in the exact sciences or not, tends naturally to think in terms of entities existing in themselves (atoms, electrical or magnetic fields and so on) and possessing properties that have values that also exist in themselves (for example the value of an electric field at various points). It is very hard for the mind to avoid hastily 'reifying' the new mathematical algorithms (wave functions, possible results of measurements not yet made and the like) that it may be presented. Paradoxically the difficulty is further increased by the very vocabulary of quantum physics, which for example uses the word 'state' when only the expression 'preparation of the system' would be truly appropriate. The fact of the matter is that such spontaneous reification creates serious risk of error, whereas this risk is minimised if quantum physics is conceived as a synthesis of prediction rules expressed in mathematical form. It is of course legitimate to reflect *post factum* on possible realist interpretations of such rules, but dangerous to think that these rules can be arrived at deductively as in classical physics from preliminary qualitative descriptions of relations between entities.

2 Here too however the physicist's attitude is not quite as clear-cut or as systematic as the philosopher's. Indeed, there are those who would say that it is not as coherent . . . . This is because, as already pointed out, a number of physicists accept the idea of methodological instrumentalism and think very strongly in such terms, while still retaining the vague idea (which they refrain from spelling out in detail) of an underlying reality; the description of which would in truth depend on metaphysics, if only the latter could be taken seriously – which they question.

3 It should be noted that there are certain exceptions to this rule. Essentially they concern quantised physical quantities (discrete eigenvalue spectra). The fact that such exceptions exist does not affect the argument put forward here.

4 Popper himself seems elsewhere to have recognised the validity of arguments of this kind, for he stresses that we can never know what part of a science is falsified by a given experiment. This, he seems to suggest, is not a problem for anyone who has understood that all theories are conjectures and cannot become anything else (cf. R. Bouveresse, [13], p.53), especially since *in practice* cross-checking generally enables us to determine beyond reasonable doubt which part of our knowledge has in fact been falsified. However we must remember that what is at stake here is a point of *principle* (since it is difficult to question the effectiveness of inductive methods as a practical tool for physicists), and if we are talking of principles the observation made in the text *does* raise serious doubts about the fundamental asymmetry between verification and falsification that lies at the heart of fallibilism.

5 All great revolutions, in the arts as well as the sciences, come about in this way. Did not Monet, himself the initiator of a great revolution, impressionism, quite clearly say that he saw himself not as a revolutionary determined at all costs to do something absolutely new but simply as a painter wanting to translate faithfully, and with joy, his impressions of nature?

6 These examples are often rather crude. For example, Popper tries to convince his readers of the hazards of induction by pointing out that the food poisoning caused by eating contaminated bread that occurred in Pont-Saint-Esprit some decades ago contradicts the general law we induce from experience that bread is nourishing. Even if we are willing to consider Baconian induction as being shaken by such counter examples it is clear that the inductive methods used by physicists are in no way affected by them.

7 I shall note here only as an aside the existence of the doctrines associated with such names as Kuhn and Feyerabend, which have enjoyed quite a vogue in recent decades. Like most physicists, I am of the opinion that the value of such 'socio-epistemological' theories does not match the stir they have created. Since fallibilism had spread the idea that theories successively collapsed, it was probably inevitable that purely sociological, perhaps even irrationalist, interpretations of science should appear and enjoy the limelight for a moment. But what has been said above about the solidity of experimental facts and theoretical equations is enough to dispose, in general terms, of the thesis of successive collapse. Whether applied to theories, as the fallibilists apply it, or to paradigms, as the socio-epistemologists apply it, the thesis can – given the solidity of experimental facts and theoretical equations – only affect *interpretations*. It is therefore meaningful only

within physical realism (a matter we shall return to in Chapter 4). And as we shall see in Part II of this book, physical realism now faces insurmountable problems and should therefore preferably be left aside in connection with the matters in hand – which deprives 'sociologism' of its substance.

In this connection it should be stressed that the weak objectivity or intersubjectivity used in this book and defined on p.32 has nothing in common with the concepts socio-epistemologists appeal to (see Chapter 11, reply to question 12). For a more detailed assessment of the thesis of sociologism, see [46].

3. Border areas of instrumentalism

1 Here I shall simply provide the results without going into their justification at the technical level. When f is not equal to either 0 or 1, 'roughly equal to' should be understood to mean that, in absolute value, the difference between p and f is at most of the order of the inverse square root of the number n of the elements composing E. When $f = 1$, G. Toraldo di Francia [8] shows that p can only, on average, differ from 1 by a number lower than $1/n$. At this stage we can therefore verify that spatio-temporal induction effectively leads, as noted in the text, to the methods which are in fact used for example in particle physics to evaluate the lower limits of the mean lives or the upper limits of branching ratios relative to rare decay modes. It goes without saying that 'approximately equal to' as used in my comments on statement 2 can and must be the object of similar detailed specification.

2 The postulate of the uniformity of nature is obviously a synthetic judgement and it cannot be empirical – for then there would be a vicious circle. It must therefore be an *a priori* synthetic judgement. This makes it easier to understand the reservations the positivists of the Vienna Circle had about it.

4. Physical realism and fallibilism

1 Certain contemporary realists, such as Popper, claim to be less concerned with *concepts* than with *truth*. It is not certain however that physicists, even those who are realists, would willingly follow them here. So strange is the world that it seems very difficult to make true statements about it without having checked the domain of validity of the concepts used in formulating these statements.

5. Microrealism and non-separability

1 More exactly, it has the form of a *linear combination of products* of a function of the coordinates of P_1 by a function of the coordinates of P_2.

2 A macroscopic example may perhaps help in understanding this particular point. Let each pair of particles (P_1, P_2) be represented by a couple of cyclists. A statistical set of pairs will therefore be represented by a group, or club, of couples of cyclists. Let us suppose that within this club a game has been organised, with the following rules. One morning, at a fixed time, everyone sets off, the husbands

cycling east and the wives west. At five in the afternoon everyone stops where he (she) is at that time and works out on a map the distance he (she) has covered. In the analogy proposed, each of these items of information corresponds to a measurement (not necessarily a position) made on the particle represented by the male (female) cyclist under observation.

The numbers obtained are of course in general all different (even if only because the cyclists have not all had the same training). This circumstance corresponds, in the case of the particles, to the fact that the wave function determines the result of the measurements only in terms of probability. More important with regard to the problem that concerns us is the existence of *correlations* between the results achieved by the male cyclists on the one hand and their wives on the other. It will be no surprise for example, if the women cycling furthest happen to be married to men among the male cyclists who cover the greatest distances. This corresponds to the correlations between measurements with regard to P_1 and P_2 discussed in the text. It is clear that this correlation within couples of cyclists can easily be explained by means of hypothesis of *different initial conditions from one couple to another* (a greater or lesser degree of shared training for example), and therefore that in order to account for it there is absolutely no need to suppose that the couples in question are physically non-separable (that is to say, that they might be linked by unknown Z rays for example . . .). It is only because the normal presentations of quantum mechanics *postulate that there are no such differences between couples* (that is, between pairs of particles having the same wave function) that, in them, physical non-separability can be inferred from the (mathematical) non-separability of the wave function of the pairs. Here, since we have abandoned the postulate, the possibility of drawing such an inference no longer exists.

3 As has been said, this kind of physical realism can also be called *mathematical realism* in so far as it appears that the only reasonable hope we can have of gaining knowledge of independent reality is based on the use of concepts inspired by mathematical physics.

4 In non-relativistic physics, when dealing with systems endowed with structural features we are free to consider together all the λ corresponding to a given time. In relativistic physics it is more consistent *not* to do so. Here however we need not concern ourselves with the problem since the systems for which we have to consider the λ (in effect, the particle sources) will be essentially localised.

5 To show this, one might also consider referring to the idea of 'forces at a distance' (for instance the gravitational force Newton postulated). However here there is no point in debating whether such a force really exists (in any case it is not instantaneous), for until the notion of separability has been defined precisely, it is idle to wonder whether a particular 'force' contradicts it. On the other hand, we can see without difficulty that the existence of universal gravitation is in no way sufficient to refute the 'principle of separability' as it is formulated in a precise way below.

6 The considerations which follow here offer an explanation of separability which is complementary to that presented in ISR [2] but which is still sufficient for our purposes.

7 This kind of *intrinsic probability* corresponds to what Popper in some contexts calls a *propensity*. Consequently it will be seen that far from rejecting the notion of propensity out of hand, I let it play its part in the arguments set out here. The rest of this book, however, seeks to show that contrary to what Popper seems long to have hoped, the trick of taking propensities into account is not enough to clear away the conceptual difficulties that the very data of physics throw in the path of the supporters of physical realism.

8 At this point one might wonder about the terminology being used. In particular, why should the same word, 'separability', be used on the one hand to express a mathematical property possessed or not possessed by the wave function of a system of particles (depending on whether it is a product of factors of individual particles or a linear combination of several similar products, as shown in the first section), and on the other to designate the principle just stated?

The answer can be illustrated by reference to a system of particles P_1, P_2. In quantum mechanics, when the wave function of system P_1, P_2 is known at a given time and that function is separable in the (mathematical) sense explained above (that is, when it is in the form of a product of functions of individual particles), then the probability of subsequently obtaining a given result when making a measurement of a physical quantity pertaining to one of the particles remains unchanged, whether or not in the intervening time a measurement (of any physical quantity) has been carried out on the other, and whatever its results may be. In this particular case, therefore, the 'principle of separability' is satisfied even for a physicist who adopts the 'metaphysical' postulate that the *objective state* and the *quantum state* of the system are identical. (It is often said of such people that they 'reject the hypothesis of hidden variables', but it is more correct to say that they adopt the hypothesis that there are none.) As against that, in cases where the wave function is mathematically non-separable the probabilities just discussed are generally different. If, therefore, we decided to follow the physicist in question by adopting what we have called his 'metaphysical' postulate, we should simply have to say, without taking the analysis any further, that in such circumstances the 'principle of separability' is violated. It is therefore clear that there is a close parallel between mathematical separability and separability in the sense in which the term occurs in the formulation of the principle.

That said, however, it must be pointed out again that, as already stressed, the 'metaphysical' postulate of such a physicist cannot be justified *a priori*. That is why merely noting that there are systems with non-separable wave functions is not sufficient to establish non-separability, so that it is necessary to take Bell's inequalities in order properly to do so.

9 As an illustration of this, we can also return to the example of the cyclists used in note 2. If, for example, the arrival points for men and women are distributed in the appropriate easterly or westerly directions between 45 and 60 kilometres from the departure point, the probability, for someone who knows only the general conditions of the game, that Mrs Smith has reached 59 kilometres, and the probability that she has reached that far given that we know Mr Smith has reached 60, are different in every case in which members of the club have trained as couples. On the other hand, for someone fully aware of Mrs Smith's physical aptitude and degree of training, the probability that she will manage 59 kilometres does not depend on Mr Smith's performance.

10 This observation is chiefly of interest to specialists in mathematical physics, who still – quite reasonably and on good historical grounds – hope to see the conceptual problems raised by the development of physics cleared up 'spontaneously' as a result of the application of increasingly abstract and highly-developed mathematical techniques. Again, such a hope is in general terms quite legitimate, and in many respects attempts to develop and generalise the formalism of mathematical physics have proved rewarding. The problem of non-separability, however, is an exception. The considerations briefly summarised here indicate that the solution to it will not be found in progress centering on formal axiomatics, which means we need a more profound *philosophical* idea of reality.

11 It is impossible to list here all the experiments in this field. Those of Clauser [18], Fry [19], and Aspect [20] are among the most precise.

12 In response to earlier publications of these matters, I was surprised to receive a certain number of letters and pre-prints of articles to the general effect that since the inequalities had been violated, those propounding them had clearly been wrong, and that since the elementary rules of quantum mechanics, with which the latter ought to have been acquainted, predicted such a violation, their mistake was inexcusable. As the inequalities in question are at present the *only* available way of testing the matter, it is to be hoped that what is reported here will refute such a simplistic pronouncement, which if it were valid would in principle rule out any argument of the type '*a* entails *b*; *b* is false; *hence a* is false'.

13 The operational schema of course admits of many variants, that which in principle consists of drawing lots for the orientation $\mathbf{a}$ or $\mathbf{a}'$ of F and $\mathbf{b}$ or $\mathbf{b}'$ of G being the best known and most useful for discussing ideas. Similarly, the Bell–CHSH inequalities can be expressed not between numbers $N_{u,v}$, but between averages of products $\alpha\ \beta$, etc.

14 For the sake of clarity, it should be noted that the position of such theorists is incompatible with that adopted by those who (wrongly) maintain that quantum mechanics itself sufficiently established the violation of the principle of separability, and that consequently merely taking into account the possibility that it might be true is already an error in itself (see Note 12). This means that any objections to the arguments presented in this chapter that the two groups might formulate cannot form a coherent whole.

15 It is true that in view of flaws in apparatus none of these experiments can, with all due rigour, be seen as crucial. But this is the place to remind ourself, as Duhem stressed, that this is always the case in science.

6. Physical realism in trouble

1 It is important to stress that the meaning of the word 'measurement' as used in microphysics is not quite the same as its meaning in macroscopic physics or in everyday language. If I use a tape to measure the length of a table for example, and find it to be 55 cm, I am justified in thinking that the table was that length to within the measurement error before I measured it. This is because the table is a macrosystem and its various characteristics cannot therefore be significantly

changed by the interaction of the instrument I subjected it to in order to take the measurement. However, with a measurement carried out on a '*microscopic*' system, such as an atom or a molecule, the situation is very different. Since the measuring instrument is macroscopic, it can easily disturb the system quite considerably while the measurement is being made. Consequently, we are not in general justified in thinking that the result we obtain expresses the value the quantity measured had – within the system – immediately before the measurement operation. It is even conceivable that the quantity had no well-defined value at that time. Nevertheless, we shall continue to talk about 'the result of the measurement', although under the convention that this merely designates the number that the instrument has registered (and that we can read on the dial) when the system being studied has temporarily been made to interact with it in conformity with the rules governing the way that instrument is used.

2 For special initial states of the system being measured, these probabilities are all equal to 0 except one, which is equal to 1. For such initial states, the result of the measurement is therefore well defined from the start, which authorises us to use the customary phrase 'measuring instruments' when speaking of those involved here.

3 Of course, those who are not put off by formulae express more succinctly what has just been said in the text. They write

$$f = c_1 u_1 v_1 + c_2 u_2 v_2 + \ldots + c_n u_n v_n,$$

where f is the wave function of S and $|c_j|^2$ ($j = 1, \ldots n$) is the probability that the measurement will give the result a_j, u_j, and v_j, being respectively the wave functions of the quantum system measured and of the instrument, immediately after a measurement having yielded result a_j. However, the formulation given in the text is quite satisfactory for understanding what follows, so readers with little time for mathematical symbols need not spend it on these formulae.

4 I have shown elsewhere [17] that the use of the more elaborate formalism of 'statistical operators' does not in any way help in this matter. On this issue see also [2B].

5 The fact that what we need to consider here is not the numerical coefficient itself but the square of its modulus is not fundamental to what is being discussed in this paragraph, which is why it has not been deemed necessary to recall the definition of the latter term.

More generally, if any reader feels at this stage that the statement made here is too learned or too abstract he really need not worry. If he continues reading he will soon see that in order to understand the observations made about this statement, all he needs to grasp is its *general form*. There is no need to understand it in detail, or to be able to apply it. Only physicists have to be able to do that.

6 A function $f(x)$ is said to be 'expanded on a set $f_1(x)$, $f_2(x)$, . . . $f_k(x)$, . . . of functions' when it is written in the form of a finite or infinite sum of functions proportional to the $f_k(x)$. Here the latter must also satisfy certain conditions.

7 Many of Niels Bohr's conceptual analyses [1] involve instruments, some of them *parts* of instruments, for example sliding diaphragms. In certain cases, these are treated as quantum objects, that is to say, they are considered in the same way as

we were considering the dial displaying numbers in the 'deterministic' description a short while ago. In other cases he treated them as *classical* objects with, by definition, a single well-defined position at a particular time. It is quite clear when he uses each kind of description. The second is used only when the diaphragm is considered as being part of the instrument. As we can appreciate, this criterion makes no reference to diaphragm properties which would be objective 'in the strong sense'. Its appeal is purely to intersubjectivity.

8 Here too it has been shown that recourse to the formalism of statistical operators is no help. See [17] and [2B].

9 They can be defined in relation to a particular frame of reference, but that definition is subordinated to the choice of this frame of reference and hence depends on the experimental device. It is true that formally speaking a functional [30] defined in terms of 'space-like' surfaces can also be associated with any system *S*. Obviously it is then possible, if one so wishes, to refer to the functional as a 'state of *S*', since any attribution of names is a matter of convention. However, when measurements are assumed to be carried out on *S* at various points in space-time [30] it is not possible unambiguously to associate with these 'states' precise values of some physical quantities belonging to *S*. This means that the answer to the question of whether quantum systems can be described by specifying such values, even at times when these systems are not being observed, must be in the negative.

7. Irreversibility

1 The objection that there is a property of weak interactions which does not respect such reversibility is not relevant. This property is so specific that physicists agree it is not a possible source of macroscopic irreversibility, which is why no further mention of it will be made here.

2 In classical physics, it is usual practice to relate any given type of system to an abstract space E with $6N$ dimensions, N being the number of particles composing a system of the type in question. With each system of this type, there is thus associated a representative point in space E, the $6N$ coordinates of the point being the $3N$ space coordinates and the $3N$ components of the velocities of the particles involved. Over time, this point describes a *dynamical trajectory* in E. In E an ensemble (statistical set) of systems is described by a *cloud* of points and ρ designates the density of the cloud at each point in E.

3 The matter is in fact more complicated and delicate than this analogy suggests, but this is not the place to enter into a digression.

4 The derivation of $\tilde{\rho}$ from *classical* ideas was mentioned here, but the Brussels School also carries out the same operation starting from quantum axioms.

8. Sensible reality

1 'Sensibility' will remain a taboo word for some time yet, and is in any case burdened with a plethora of possible misinterpretations. For that reason, from now

on I intend to avoid it and instead refer to the 'right hemisphere' of the brain and its activities. It should however be noted that in view of present knowledge to do so would border on misinformation were it not for the fact that I here make it quite clear that I adopt absolutely no position on the question of where the 'true seat' of the emotions is located, or even on the meaning of such a question. The references to left and right hemispheres in this book are therefore to be understood in an essentially metaphorical sense.

9. Independent reality

1 If it is permissible to abbreviate and simplify a history of ideas whose riches are found chiefly in its subtleties, we might present things according to the following schema. In the seventeenth century, the idea that subject and object are separate came to the fore. From then on, almost to the present day, science developed on the premise that it is possible to treat objects as things in themselves. Parallel to this development in science, however, some philosophers came increasingly to feel that they had to recognise in the very elements of discourse – space, time and causality – the *a priori* forms of human sensibility and human understanding. An almost necessary corollary of this view was the convention that things in themselves are not accessible to human experience, which in turn gave rise to the view that the idea of things in themselves should be rejected. Accordingly, in the nineteenth and twentieth centuries numerous philosophers came to deny that both the idea of things in themselves, and (hence) 'independent reality', had any meaning.

From the start however, those who thought in this way had to face the obvious difficulty it raised: that such a thesis would inevitably end in nothing but solipsism, pure and simple. Aware of this pitfall, many of them introduced various notions to avoid it (transcendental consciousness, the cosmic subject, and so on). It would not be putting it too strongly, in one sense at least, that to a large extent the extreme complexity of philosophical enquiry in our day is the inevitable consequence of the intractable nature of the problem that thus arose.

If we sought to be rigorous, the rules of good practice would oblige us to examine in systematic fashion all these diverse attempts; it would be helpful if one day the rather particular scrutinising style of a specialist in the exact sciences were systematically brought to bear on this burgeoning forest of systems. Here though there is no question of putting the reader to such an endurance test, which in any case would demand, by comparison with what has gone before, a radical change in the mode of reasoning used in this book. So, without going into sufficient detail to provide a proper justification for my claim, I shall simply put forward the idea that from many points of view the various notions introduced by philosophers in an effort to overcome the problem of solipsism – transcendental ego, undifferentiated consciousness, and the like – can all be interpreted as unacknowledged reversions to the idea of some kind of independent reality. This seems particularly true with regard to those philosophers who explicitly reject any portrayal of them as closely or distantly akin to idealism or to 'spiritualist' doctrines.

2 There is an oft-told tale that when he was a young man Boltzmann discovered to his great chagrin that a book on physics said to be totally clear and to provide quite rigorous definitions of all the terms it used was written in a language (English) which he did not understand. His brother, it is said, set him laughing with the

'common sense' reply that if the book was truly as described, reading it would not pose Boltzmann any problems, for he would find the meaning of each word was clearly defined . . . .

3 Some authors understand the 'ergo' in *cogito ergo sum* as an expression of *causal* necessity. This would mean that 'cogito ergo sum' – I think therefore I am – would be rather like assertions of the kind 'He has not worked, therefore he will fail his examinations'. The point I am discussing here relates to the *cogito* only if Descartes' claim is understood differently; more precisely, as meaning 'If I were not, I could not think; but I do think, hence it is clear that I am' – that is, as expressing a *logical* necessity (we should perhaps then prefer the translation 'I think, hence I am', which can more easily be understood that way).

4 Without in any way wishing to poach on philosophical preserves, I must point out that the classical objections to the argument of the *cogito* deal with other matters than the one under discussion here. Essentially, they consist in noting that on the one hand the argument does not demonstrate the existence of a personal 'I' endowed with temporality and memory, and on the other, that concerning the question of justifying the assertion 'something is', the facts of being conscious of oneself and of the existence of objects should probably be granted the same status. However, here we are concerned with neither 'self' nor 'objects' but simply with the indubitability of the existence of at least one thing (or one attribute of 'something'). And this certainty we do indeed have concerning consciousness as such. Whether it be of an individual self, of objects, or both together, is not what is important at this level.

5 This argument was notably put forward by the physicist Rudolph Peierls during the discussion which followed a lecture by the author.

6 Jules Vuillemin, private communication (I thank Prof. Vuillemin for aiding me to define the problem more clearly).

7 Ewing's short book *The Fundamental Questions of Philosophy* [39] summarises clearly, precisely and objectively the arguments for the various theses and offers a useful first introduction to the problems we are discussing.

8 Such is the substance of Popper's response when his realism is attacked. My own realism is of a very different kind, but I endorse the answer.

9 Here we are not taking into account the positions taken by one or two rare physicists who claim with equal vigour *both* that the proper field of physics is very restricted *and* that the theory of the relativity of states (see Chapter 6) ultimately is *the* correct theory. Since the latter explicitly postulates that the whole Universe is strictly subject to the laws of quantum physics (universal wave theory), their position is clearly self-contradictory.

10 This description of the four theses was directly inspired by a lecture given by Dr G. Rager to the members of the International Academy of the Philosophy of Science (Brussels, April 1984).

10. The dilemma of modern physics

1 A first version of the contents of this chapter appeared in *Quantum Implications, Essays in honour of David Bohm* [41]. Here, while I have avoided technical details as far as possible, where they were not indispensable, I have edited the text on the principle that it would mostly be read by physicists.

2 The study presented in this paragraph does not appear in Wheeler's writings. I offer it here because I think it removes a difficulty.

3 When the instrument is treated like a quantum system, the evolution of the system *A plus P* during the interaction between *A* and *P* which precedes making the observation considered here should be calculated by studying the evolution over time of the complete wave function of the system *A plus P*. One will then find that for some kinds of initial conditions this wave function cannot, once the interaction in question is over, be put into the form of a product of a wave function relating to the particle and a wave function of the instrument corresponding to a well-defined value of the position of the needle on the dial. We have already looked at this in Chapter 6.

4 To make the paradox more striking, Schrödinger [45] replaced the needle by a cat which either died from something triggered by the system, or survived.

5 A quantum state is said to be an eigenstate of some physical quantity if the probability that, upon measurement, this quantity is found to have some particular value is equal to one for one of the possible values, and zero for the others.

6 Physicists call 'reduction' (or 'collapse') of a wave function (or quantum state) the fact of a measurement putting the system into an eigenstate of the measured quantity.

7 As we have seen, the difficulty can be solved *formally* by identifying the state of a system with a 'functional' [30]. However, this is only a precise way of saying that it is impossible to consider the physical state of the system as something existing by itself, independently of the observations that might be made at various points in space-time. Such a resolution is therefore of a *purely formal* nature and does not in any way reduce the conceptual difficulty. For more details, see Chapter 6, note 8.

8 Incidentally, the content of this section makes it possible to answer a question that has nothing directly to do with the matters in hand but is interesting in itself. It is the question of assessing the value of the claim that *any* measurement (by which I mean any human action directed to measuring by means of an instrument a physical quantity pertaining to some object) would necessarily disturb the state of the system being measured. Since the claim is sometimes too readily conceived as expressing a general feature of microphysics, it must be pointed out that when taken literally it is false not only in classical physics but also in 'conventional' quantum physics. In classical physics, if for example I know that a correlation has already been established between two systems *U* and *V* at present separate, I can increase my knowledge of *U* without in any way disturbing it simply by measuring an appropriate physical quantity pertaining to *V*. In elementary quantum mech-

anics there is a whole class of measurements which, at least if we accept the basic formalism of the subject, can be carried out without in any way disturbing the quantum state of the system measured. These measurements involve observables of which the state in question is already an eigenstate.

No doubt it is in confusing the conventional approach (which is, to repeat, the one that is used in practice) and the approach which uses partial definitions (and was adopted by Bohr in his refutation of the argument of EPR) that the origin of the claim under discussion is to be found. It is clear, however, that even if the second approach is adopted the claim is misleading and inappropriate, at least as formulated. Obviously, a measurement cannot disturb a microscopic quantum state which, if we adopt that approach, does not exist (and we know that Bohr, for just that reason, refused to use the expression 'disturbing phenomena by observation' [1]).

11. Questions and anwsers

1 This procedure has not only the advantage of being more readable. In many cases, it also results in greater accuracy. Most of the diverse observations and ideas I shall examine were addressed to me on the publication of one of my earlier books, *In Search of Reality* [2], where many of the problems investigated in the present work were also examined (albeit from an angle that was often somewhat different). The value for an author of such questions, objections and observations can, as it may be imagined, be very great indeed, and I am extremely grateful to the well-informed people who put them to me.

2 It will be seen that there is – beyond the agreement always possible on points of detail or principles of action – a genuine conflict which it would be idle to conceal between the intellectual starting point of many philosophers compared to that of many scientists. The refusal I have just explained reflects, it seems to me, an attitude common to most of the latter. A fair number of philosophers, on the other hand, seem from the beginning to have adopted the inverse attitude, perhaps because of the training they have received. The following remark, taken from L. Brunschwicg's *Ecrits philosophiques* [47] (and which, I hasten to add, should not be taken as typical of his rather subtle thinking), illustrates this appropriately: 'a philosophy of science is necessarily idealist because it is obvious that the object of science is constituted progressively, bound up as it is with the network connections that the intelligence of man establishes ever more extensively and ever more densely in the field of space-time'. It is clear that in respect of the connection of facts and consequences, what was obvious to Brunschwicg was not obvious to Einstein, for Einstein – who was clearly as aware as anyone of the fact to which Brunshwicg alludes, namely (as seen from the context from which the above quotation is drawn) the conceptual transformation wrought by relativity – nevertheless continued to declare himself a realist.

Certainly, we could try to reduce this difference to one of vocabulary alone. The idea that what Brunschwicg calls idealism is not as opposed as it seems to what I, with Einstein, have called realism, could be defended. But to my mind such an assimilation would be rather forced, since it would tend to leave out of account the obvious difference, at least of emphasis, between what each judges to be most significant (in respect of basic ideas, of course). In fact, the reason for this difference, in my opinion, is that the philosophers in question (and Brunschwicg in particular) too hastily mixed up 'mathematism' and idealism. They have not fully grasped the logical possibility opened by the notion of 'distant realism', or Pythagorism, which rests on the idea that it is conceivable that the *true* concepts, those which correspond to the true structure of the real in itself, may be initially unknown but ultimately knowable, that is to say, that mathematical physics can in the end lead to their discovery. Such a hope, which they would perhaps regard as illusory, was certainly held by Einstein.

3 The hypothesis that the mind of the individual observer might play an active part in the *reduction of the wave packet* has been envisaged favourably – but only as a hypothesis – by a number of theoretical physicists, including some who have made major contributions to science. This in itself is enough to establish that the hypothesis – even if, ultimately, it proves more satisfactory to replace it with a sharp distinction between independent and empirical reality – was inspired by certain objective facts and hence is not pure fantasy. Setting this very special case aside, it needs to be stressed that the explicit introduction of the notion of mind into the precise and detailed descriptions of the physics of phenomena cannot at present be founded on any scientific argument weighty enough to resist objective criticism of the kind traditionally deployed in discussing hypotheses in the sciences. This rules out as *scientifically non-valid* works centred on the notion of mind and claiming to provide thereby new and detailed descriptions of a number of physical phenomena. Generally speaking, the precision and abundance of detail in such descriptions, and the fact that they are presented as physical discoveries in the proper sense are in this matter strongly negative indicators which should arouse the critical instincts of the reader.

4 It also remains that the remark of Jacques de Bourbon-Busset that 'desire is the master' is in the final analysis profoundly true of this as of all other fields.

5 Nor is it absurd to wonder whether Kant himself did not at certain moments harbour somewhat similar ideas about things in themselves. His celebrated comparison between Reason in physics and a judge who 'forces the witnesses to respond to the questions *he* poses them' is usually quite rightly cited as an illustration of the Kantian idea that reason sees only what it produces itself after its own plans. But it is still no less true that while the judge may be master of the questions he poses, he is not master of the answers. The nature of these answers is at least in part determined by what has happened outside his tribunal. Moreover, it is just that, what has happened outside his tribunal, to which the judge is anything but indifferent. However obscure and partly inaccessible the truth of the matter at hand may be, his goal, which he does not take to be illusory, is to try to establish it as best he can, and *that* is what is important to him.

6 I have discussed this point to some extent elsewhere. See *Un atome de sagesse* [46].

7 'We must call *science* only *the collection of recipes that always succeed*. All the rest is literature' (Paul Valéry). It should be noted that in other more finely shaded passages Valéry raises questions which anticipate somewhat those appearing here.

8 In a sense this confirms certain intuitions of Niels Bohr, since he refused to make the wave function (the 'orbitals of the chemist) an element of reality.

9 'Difficult' is exactly the right word here. If there is one field in which one should not content oneself with incisive arguments then surely it is this one. And physicists do *not* so content themselves. Several research groups are currently studying, from this point of view, some particularly interesting macroscopic systems. These (for example, confined magnetic fields in hollow superconductors) are of the kind in which certain specifically quantum properties (such as quantum superpositions of states that are macroscopically separated in space) cannot, if they exist, be entirely veiled by the effects of 'thermal agitation' or of non-isolation, whereas, according to the theory itself, they unavoidably are in most kinds of macroscopic systems. Though such research is still underway, and though in this domain experimental testing is still undertaken at the extreme limits of what is technically possible, it can now be said that all the indications are that the behaviour of such systems conforms in detail to what quantum mechanics predicts.

10 With regard to the proofs of the fact that objectivity of both quantum state and of causality is merely weak, the initiated reader may consult either *Nonseparability and the Tentative Descriptions of Reality* [2A], or for greater detail, the original articles quoted in it, in particular those of Aharanov and Albert [28], [29], and Srinivas [50].

11 Admittedly, the quantum framework has not, as one might well think, worn the same 'clothes' for the whole half-century of its existence. It has been greatly modernised in form and, it might be said, generalised in substance. A rather small minority of theoretical physicists – composed in part of those who contributed to retailoring of its clothes – seem to think that such developments should allow of an 'objective' solution to the problems addressed in this book (the theory of the measurement operation, quantum reduction and relativity, and so on). Unfortunately, these workers are most of the time too exclusively concerned with formal questions of pure mathematics for it to be easy for them to become aware of the truly conceptual aspects of the problems they seek to solve. Many theories of the measurement operation, for example, are presented as providing an objective solution, without any real indication of what nevertheless becomes apparent once the details of the solution are scrutinised: that the solution presented is only weakly objective, in the sense of the term I have taken care to explain.

12 Physical theories of 'internal symmetry' and their recent extensions, the celebrated 'gauge theories' deserve special mention. They are based on a kind of generalised geometry, and those physicists who, like the present author, took part at some time in developing and elaborating particular aspects of the larger whole they constitute are generally – and quite rightly – impressed by their great synthesising power. While guarding against undue optimism (for the work is still in progress and the difficulties encountered would have to be described) many of these physicists are of the opinion that, thanks to such theories, we shall soon be able to look upon all the dynamics of elementary particles and of fields as deriving from an invariance principle of the physical laws under gauge transformations. This has tempted some to elevate the principle to the status of a new theoretical framework, and to claim that with the arrival of the super-geometry alluded to,

Einstein's hopes of constructing a theory that would be compatible with mathematical realism are on the way to being realised. My personal view, however, is that when presented in that manner, the idea disguises a certain confusion. It is perfectly legitimate, I think, to believe that the *classical* aspects of gauge theories would have aroused Einstein's enthusiasm, for exactly the reasons just given. But we must not forget that if such theories are to represent phenomena accurately they must be quantised. It is at this level that the limited and purely technical progress that has been made has in no way solved the conceptual problems which were posed before the appearance of the theories in question. It therefore seems quite clear that Einstein would not see these theories as realisations of his ideal (and the same goes for the present-day 'Einsteinians' working on them). In the circumstances then, it is not under-estimating the great importance of internal symmetry and gauge theories to refuse to see them as components of a new theoretical framework to replace the quantum framework.

12. Summary and perspectives

1 It is well known that many positivists think they can avoid such extreme views by accepting verifiability of a merely logical kind. In particular they maintain that reference to assertions about the past can be made meaningful by use of such propositions as 'if we lived in the Mesozoic era, we would see diplodocuses'. However such a procedure leads to all the difficulties associated with counterfactual statements and definitions by counterfactual conditions.

2 It would be too much of a digression here to dwell on the absolute necessity of such an acceptance if we are ever to see the kind of society we live one day return to its senses. This indeed is so important that to merely state it, in the measured terms appropriate to analysing ideas, should not be considered adequate. It should be shouted from the rooftops, proclaimed between television shows – or better, in place of them – on every network in the world. There was a time when what we are pleased to call 'advanced' societies thought they could make progress by rejecting any such aspiration. They thought they could reduce the whole of Being to phenomena, and phenomena to action. But (as we have seen with regard to positivism) that was an illusion. Paradoxically, just at the time when the best minds thought that this move had purified our intellects it became apparent that it was intellectually corrupt and no more than a mirage. Alas it is a mirage which though it deceives scarcely any of the genuine thinkers of these last decades of our century, still holds sway over most of the human beings who have a platform for their ideas. Consequently its hold on the collective unconscious of the masses in the West is undiminished. If this book were more practical in its orientation (one cannot deal with everything, least of all in a single book) this theme would receive the elaboration it deserves. With all necessary illustration and qualification, it would denounce the three kinds of propaganda – propaganda of frenzy, propaganda of violence and propaganda of manichaeism – that the 'communicators' of our part of the world vie with each other to produce (those of the Eastern bloc and the Third World have their blemishes too, but we cannot do much about them). It would explain why the development of all three is a direct consequence of the denial of any reality external to human beings and hence of the refusal to think about such reality. Correlatively it would of course also include appropriate analyses of the all

too numerous perversions of this 'attention to the real', together with studies of ways to avoid them and their negative consequences. Alas such complex questions – which, among others, I sought to tackle elsewhere (see *Un atome de sagesse* [46]) – would take us too far from our theme here. That theme, to repeat, bears essentially on questions of knowledge, even if, as we shall see later, that term has to be taken in a broad sense.

3 This condition in no way rules our personal convictions. However it does oblige anyone who holds them to abjure any idea of a crusade and *a fortiori* of a holy war. Indeed we should say more: it forbids us from transforming such a personal conviction into a feeling of hostility to anyone whose personal convictions are different from our own. In view of the constant participation (much to be desired) of the 'right hemisphere of our brain' in such convictions, this involves an intellectual asceticism of an undeniably difficult kind. And that, no doubt, is where the rational aspect of the activity in question must come to exert a positive influence.

4 Some phenomena concerning such molecules provide an exception, in that studying them requires that fundamental (that is to say, quantum) laws be taken into account. However these are sufficiently few that rightly or wrongly a conceptual description which essentially ignores them is generally deemed acceptable.

5 Even this is highly schematic. In fact specialists in artificial intelligence, the functioning of the brain and the like long ago discovered that the hypothesis of a strict idea-neurone correspondence is untenable. As Dennett [52] notes with some humour, if for example the representation of the concept of 'grandmother' was a matter for a particular neurone or group of neurones, their natural disappearance in an otherwise healthy person would give rise to the selective impossibility of him recognising his grandmother, or more radically, of even thinking the idea 'grandmother' – an affliction which does not seem to be observed. Hence such specialists are obliged to seek a more abstract correspondence. But most of them – no doubt taking as their starting point the fact that specifically quantum effects seem *in practice* to be of little importance in this context – find it quite natural to look for such correspondences within an overall conception which they call 'physicalist'. According to Dennett, such a conception is centred on the basic idea that 'every mental event is (identical with) a physical event in the brain', and that consequently 'we don't need a category of non-physical things in order to account for mentality'.

While acknowledging the lucidity and impartiality with which such specialists conduct their search (which they admit is not easy), I believe it is necessary to stress a logical point which seems to make difficulties here. It turns on the following fact, signalled in the text. Under pain of circularity the basic idea mentioned above can *a priori* constitute the framework of a description of mental phenomena only if the concept of a physical event is itself defined first, and moreover is defined (at least as far as the events under consideration are concerned) independently, even in principle, of notions such as 'measurement', 'observational conditions' and so on, because obviously these notions themselves have to do with the more general notion of 'mental phenomena' which is precisely what is to be cleared up. Given the facts as reported in this book, it seems to me rather rash to assume this necessary condition has been met. This prompts me, not of course to deny the intrinsic interest of detailed exploration of the relationship between thinking and the brain (Dennett's book is just one of a number of studies which brilliantly illustrate the wealth such exploration can provide), but rather to have considerable reservations about any position which claims to *conclude* from the results of such research that the 'physicalist reduction' is valid. The situation is at very least more delicate – and philosophically less explored – than any such claim would have us believe.

6 The expression 'could quite well be' always served to indicate that for the person using it, what it refers to remains in the realm of mere conjectures. This is just how it should be taken here. More generally, the reader should note that after the present paragraph we enter the realm of personal conjecture. There can be no genuine conflict of ideas between the author and anyone who rejects the content of these closing pages but accepts that of the rest of the book.

7 It is not always sufficiently noted that in Dante's time the idea educated people had of God was much less that of a creator, or of a 'watchmaker' (though of course this idea was abroad, it was not at the forefront) than that of a final cause (by which is meant a God such that the highest end of man and of the world was 'one day' to attain unity with Him for all eternity). No doubt it is to this latter idea, with all its inspirational power for the inner life, that we owe for example the treasures of early Italian art. And it was not until after Galileo that the spread of the mechanistic view of reality made such an idea paradoxical; thenceforth thinkers had to content themselves with a conception previously held, according to Burtt in his *Metaphysical Foundations of Modern Physical Science* [53], to be cruder and more popularistic, namely that of God as creator. If such thinkers were not to be false to what they considered they knew, they could attribute to God scarcely any role other than to have given the initial 'push' to the Universe. But it is clear that not even the theological zeal of a Newton or a Maxwell could confer on this new conception of God as essentially a 'watchmaker' the character of a vision which could be for the spirit as animating as the conception it displaced.

8 I sincerely hope that after scanning the last few pages rapid-reading enthusiasts will not be carried away with the idea that I am claiming to use physics as a means of re-instating the idea of final cause in the normal sense, that is to say, as acting among phenomena. They should be quite clear that the distinction between independent and empirical reality is absolutely essential to the coherence of what I have just been discussing and that I certainly do not believe in the existence of final causes that operate within the domain of empirical reality. What I have said is in no way to be passed off as down-market occultism, fairground mysticism or anything of that ilk. In the culture of our day there are strong tendencies driving us to think in that way, but these tendencies are perverted.

9 Readers wishing to do so should first re-read note 3 of Chapter 11, which deals with such fable makers.

10 In general, the right hemisphere is more demanding. It needs to be able to give itself enthusiastically to beautiful things. Now if things are to be beautiful they must have a shape . . . . In order that this desire for 'precise shape' should not lead a great number of people with fresh and open minds to fall for excessively childish beliefs, no doubt it is well to accept the notion of the 'imaginary' as a category

intermediate between the 'true' and the 'false'. Like the 'model' of the scientist, the 'imaginary' has something of the parable about it. But it is a parable in the literal sense, in which the individual reserves some kind of right to believe even though he knows that 'everything is, of course, rather more complicated'. For my part, I would go so far as to extend my sympathies to those gurus, shamans and leaders of sects who are honest enough to admit that it is within the framework of such a realm of the imaginary that they operate . . . .

Appendix 1

1 Popper compares the intersubjective agreement in question here to 'the verdict of a jury'. As I see it, the analogy is significant on condition that we understand by 'verdict' not a decision about the fate of the accused (be sentenced, acquitted, etc.) but about the statement of a strong belief (innocent, guilty, . . .) about the facts observed (directly or by a third person) shared by the majority. Then the difference between the two kinds of judgement is indeed one of degree only. But this difference of degree is nonetheless very important. The majority which forms among observers on whether a particular signal light is on or off will always be potentially infinitely larger than that which has a chance of forming in a supposedly arbitrarily large courtroom jury, even in the simplest cases. We might almost go so far as to say that here is an example of a situation in which a very important quantitative difference amounts to a qualitative difference.

Addendum

1 Should we here take advantage of the presence in our languages of the two verbs 'to be' and 'to exist', and use one of them as a synonym for 'to be' in the strong sense, while using the other one as a synonym for 'to be' in the weak sense? Some philosophers do make a distinction between the meanings of these two verbs and we could of course follow their example. Whether a common standpoint would thereby be reached between them and us remains however a debatable question.

2 In the present author's opinion the use some theorists – in particular Primas [72, 73] – make of such words and expressions as 'to be', 'ontic states' and so on can be reconciled with other statements made by them only if, in their writings, the verb 'to be' is given this weakened, human-centred sense.

References

[1] N. Bohr, *Atomic Physics and Human Knowledge*, Science Editions, New York (1968).

[2] B. d'Espagnat, *In Search of Reality*, Springer Verlag, New York (1983) (ISR); *Nonseparability and the Tentative Descriptions of Reality, Physics Reports, 110,* no.4, p. 202-264, North-Holland, Amsterdam (1984), (ref. 2A). *Conceptual Foundations of Quantum Mechanics* (second ed. 1976), Addison-Wesley-Benjamin-Cummings, Reading, Mass., (ref. 2B).

[3] K. Popper, *Unended Quest: An Intellectual Biography*, Fontana, London (1976).

[4] W. Heisenberg, *Physics and Beyond*, Allen and Unwin, London (1971).

[5] C. Chevalley, *Mécanique quantique et positivisme*, 1925–1939 (dissertation).

[6] R. Carnap, *Meaning and Necessity*, Phœnix Books, Canaan, NH (1958).

[7] M. Clavelin, *Les deux positivismes du Cercle de Vienne*, Archives de philosophie, **43**, p. 33 (1980).

[8] G. Toraldo di Francia, *The Investigation of the Physical World*, Cambridge University Press, Cambridge (1981).

[9] N.R. Hanson, *Observation and Interpretation in Philosophy of Science Today*, S. Morgenbesser ed., Basic Books, New York (1967).

[10] K. Popper, *Logik der Forschung, Springer, Wien* (1934).

[11] G. Radzintsky, *Revue internationale de philosophie*, **131-132**, p. 179 (1980).

[12] *The Philosophy of Karl Popper*, P. Schilpp ed., Library of Living Philosophers, La Salle, Ill. (1974).

[13] R. Bouveresse, *Karl Popper*, Vrin, Paris (1981).

[14] C.G. Hempel, *Methods of Concept Formation Science*, International Encyclopedia of United Science, University of Chicago Press, Chicago, Ill. (1953).

[15] R. Carnap, *The Foundations of Physics*, Basic Book, New York (1966). Armand Colin, Paris (1973).

[16] D. Lewis, *Counterfactuals*, Library of Philosophy and Logic, Basil Blackwell, Oxford (1973).

[17] B. d'Espagnat, *Conceptions de la physique contemporaine*, Hermann, Paris (1965);
An Elementary Note about Mixtures, in *Preludes in Theoretical Physics*, A De Shalit *et al.* eds., North-Holland, Amsterdam (1966).

[18] S.J. Freedman and J.F. Clauser, *Phys. Rev. Letters*, **28**, p. 938 (1972).

[19] E.S. Fry and R.C. Thompson, *Phys. Rev. Letters*, **37**, p. 465 (1976).

[20] A. Aspect, P. Grangier et G. Roger, *Phys. Rev. Letters*, **49**, p. 91 (1982);
A. Aspect, J. Dalibard et G. Roger, *Phys. Rev. Letters*, **49**, p. 1804 (1982).

[21] B. d'Espagnat, *The Quantum Theory and Reality, Scientific American*, **241**, p. 158 (1979).

[22] R. Thom, Halte au hasard, silence au bruit, *Le Débat* (Gallimard), **3**, p. 119 (1980).

[23] I. Lakatos, Falsification and the Methodology of Scientific Research Programmes, in *Criticism and the Growth of Knowledge*, I. Lakatos and A. Musgrave eds., Cambridge University Press, Cambridge (1971).

[24] L. de Broglie, *Journal de physique*, **5**, p. 225 (1927).
[25] D. Bohm, *Phys. Rev.*, **85**, p. 166 et 180 (1952).
[26] H. Everett, *Rev. Mod. Phys.*, **29**, p. 454 (1957).
[27] *Quantum Theory and Measurement*, J.A. Wheeler and W.H. Zurek eds., Princeton University Press, Princeton, N.J. (1983).
[28] Y. Aharonov and D.Z. Albert, *Phys. Rev.*, **D21**, p. 3316 (1980).
[29] Y. Aharonov and D.Z. Albert, *Phys. Rev.*, **D24**, p. 359 (1981).
[30] Y. Aharonov and D.Z. Albert, *Phys. Rev.*, **D29**, p. 228 (1984).
[31] D. Bohm, *Wholeness and Implicate Order*, Routledge and Kegan Paul, London (1980).
[32] I. Prigogine et I. Stengers, *Order out of Chaos*, Bantam, New York (1984).
[33] I. Prigogine, *Physique, temps et devenir*, Masson, Paris (1980).
[34] A. Peres, *Phys. Rev.*, **D22** p. 879 (1980).
[35] E.P. Wigner, in *Proceedings of the International School of Physics*, Enrico Fermi (Varenna), Course 49, B. d'Espagnat ed., Academic Press, New York (1971).
[36] H.D. Zeh, *Foundations of Physics, 1,* p. 67 (1970).
[37] H. Barreau, *Bergson et Einstein*, Les études bergsoniennes, X, p. 90, P.U.F., Paris (1973).
[38] R.P. Feynman, *The Character of Physical Law*, The M.I.T. Press, Cambridge, Mass. (1965).
[39] A.C. Ewing, *The Fundamental Questions of Philosophy*, Collier, New York (1962).
[40] J.A. Wheeler, Bits, Quanta, Meaning in *Problems in Theoretical Physics*, A. Giovannini, F. Mancini and M. Marinaro eds., University of Salerno Press (1984).
[41] J.S. Bell, Beable in Quantum Theory, in *Quantum Implications*, B.J. Healey and F.D. Peats eds., Routledge and Kegan Paul, London (1987).
[42] W. Heisenberg, *Physics and Philosophy*, Allen and Unwin, London (1971).
[43] A. Einstein, B. Podolsky et N. Rosen, *Phys. Rev.,* **47**, p. 777 (1935).
[44] N. Bohr, *Phys. Rev.,* **48**, p. 696 (1935).
[45] E. Schrödinger, *Naturwissenschaften*, **23**, p. 807 (1935).
[46] B. d'Espagnat, *Un atome de sagesse*, Seuil (1982).
[47] L. Brunschwicg, *Ecrits philosophiques*, P.U.F. (1958).
[48] P. Valéry, *Tel quel*, Gallimard (1924–44).
[49] G. Hottois, *Revue de théologie et de philosophie*, **114**, p. 121 (1982).
[50] M.D. Srinivas, *J. Math. Phys.,* **20**, p. 1593 (1979).
[51] E. Agazzi, *Science et foi*, Massimo, Milan (1983).
[52] D.C. Dennett, *Brainstorms*, The M.I.T. Press, Cambridge, Mass. (1981).
[53] E.A Burtt, *The Metaphysical Foundations of Modern Physical Science*, Routledge and Kegan Paul, London (1924).
[54] J. Beaufret, *Entretiens*, P.U.F. (1984).
[55] M.B. Hesse, *The Structure of Scientific Inference*, University of California Press (1974).
[56] W.V. Quine, *From a Logical Point of View*, Harvard University Press, Cambridge, Mass. (1953).
[57] A Shimony in *Proc. of Int. Symp. Foundations of Quantum Mechanics* (Physical Society of Japan, 1983).
[58] J. Jarrett, unpublished thesis, University of Chicago (1983). C. Bastide, *Phys. Letters,* **103 A**, p. 391 (1984).

[59] O. Costa de Beauregard, Time, The Physical Magnitude, *Boston Studies in the Philosophy of Science*, D. Reidel Publ. Co. Dordrecht (1987).
[60] S. Ortoli et J.P. Pharabod, *Le cantique des quantiques*, éd. La découverte, Paris (1984).
[61] O. Costa de Beauregard, *Proc. of the Int. Symp. on the Foundations of Quantum Mechanics*, Tokyo (1983).
[62] F. Bonsack, Esquisse d'un réalisme non métaphysique, *Lettres épistémologiques, 31*, p. 3 (1981).
[63] J.A. Wheeler, in *Mathematical Foundations of Quantum Theory*, ed. A.R. Marlow, Academic Press, New York (1978).
[64] S. Machida and M. Namiki, *Prog Theor. Phys.*, **63**, 1457, 1833 (1980).
[65] S. Machida and M. Namiki, *Proceedings of the Internal Symposium on Foundations of Quantum Mechanics (ISQM)* in Tokyo 1983, p. 127, 136.
[66] H. Araki, *Prog. Theor. Phys.*, **64**, 719 (1980); *A Continuous Superselection Rule as a Model of Classical Measuring Apparatus in Quantum Mechanics*, talk given at the conference on "Fundamental Aspects of Quantum Theory" at Como, Italy, September 2–7, 1985.
[67] W.H. Zürek, *Phys. Rev.*, **D24**, 1516 (1981); **D26**, 1862 (1982).
[68] H.D. Zeh, *Found. Phys.*, **9**, 803 (1979).
[69] K. Hepp, *Helv. Physica Acta*, **45**, 237 (1972).
[70] J.S. Bell, *Helv. Physica Acta*, **48**, 93 (1975).
[71] C.M. Lockhart and B. Misra, *Physica A*, **136A**, 47 (1987).
[72] H. Primas, *Lecture Notes in Chemistry*, Springer (1981).
[73] H. Primas, Foundations of Theoretical Chemistry, in *Lecture Notes for the NATO Advanced Study Institute*, R.G. Wooley ed. Plenum Press, New York (1980).

Index